T0336931

The Riemann Boundary Problem on Riemann Surfaces

Mathematics and Its Applications (*Soviet Series)*

Managing Editor:

M. HAZEWINKEL
Centre for Mathematics and Computer Science, Amsterdam, The Netherlands

Editorial Board:

A. A. KIRILLOV, *MGU, Moscow, U.S.S.R.*
Yu. I. MANIN, *Steklov Institute of Mathematics, Moscow, U.S.S.R.*
N. N. MOISEEV, *Computing Centre, Academy of Sciences, Moscow, U.S.S.R.*
S. P. NOVIKOV, *Landau Institute of Theoretical Physics, Moscow, U.S.S.R.*
M. C. POLYVANOV, *Steklov Institute of Mathematics, Moscow, U.S.S.R.*
Yu. A. ROZANOV, *Steklov Institute of Mathematics, Moscow, U.S.S.R.*

Yu. L. Rodin
Institute of Solid State Physics,
Academy of Sciences of the U.S.S.R., Moscow, U.S.S.R.

The Riemann Boundary Problem on Riemann Surfaces

D. Reidel Publishing Company
A MEMBER OF THE KLUWER ACADEMIC PUBLISHERS GROUP
Dordrecht / Boston / Lancaster / Tokyo

Library of Congress Cataloging in Publication Data

CIP

Rodin, Yuriĭ Leonidovich, 1936–
The Riemann boundary problem on Riemann surfaces / by Yu. L. Rodin.
p. cm. — (Mathematics and its applications (Soviet Series))
Bibliography: p.
Includes index.
ISBN 90-277-2653-1
1. Riemann surfaces. 2. Riemann-Hilbert problems. I. Title.
II. Series: Mathematics and its applications (D. Reidel Publishing Company).
Soviet series.
QA333.R64 1987
515'.223—dc 19 87-28869
CIP

Published by D. Reidel Publishing Company,
P.O. Box 17, 3300 AA Dordrecht, Holland.

Sold and distributed in the U.S.A. and Canada
by Kluwer Academic Publishers,
101 Philip Drive, Norwell, MA 02061, U.S.A.

In all other countries, sold and distributed
by Kluwer Academic Publishers Group,
P.O. Box 322, 3300 AH Dordrecht, Holland.

All Rights Reserved
© 1988 by D. Reidel Publishing Company, Dordrecht, Holland
No part of the material protected by this copyright notice may be reproduced or
utilized in any form or by any means, electronic or mechanical
including photocopying, recording or by any information storage and
retrieval system, without written permission from the copyright owner

Printed in The Netherlands

SERIES EDITOR'S PREFACE

Approach your problems from the right end and begin with the answers. Then one day, perhaps you will find the final question.

'The Hermit Clad in Crane Feathers' in R. van Gulik's *The Chinese Maze Murders*.

It isn't that they can't see the solution. It is that they can't see the problem.

G.K. Chesterton. *The Scandal of Father Brown* 'The point of a Pin'.

Growing specialization and diversification have brought a host of monographs and textbooks on increasingly specialized topics. However, the "tree" of knowledge of mathematics and related fields does not grow only by putting forth new branches. It also happens, quite often in fact, that branches which were thought to be completely disparate are suddenly seen to be related.

Further, the kind and level of sophistication of mathematics applied in various sciences has changed drastically in recent years: measure theory is used (non-trivially) in regional and theoretical economics; algebraic geometry interacts with physics; the Minkowsky lemma, coding theory and the structure of water meet one another in packing and covering theory; quantum fields, crystal defects and mathematical programming profit from homotopy theory; Lie algebras are relevant to filtering; and prediction and electrical engineering can use Stein spaces. And in addition to this there are such new emerging subdisciplines as "experimental mathematics", "CFD", "completely integrable systems", "chaos, synergetics and large-scale order", which are almost impossible to fit into the existing classification schemes. They draw upon widely different sections of mathematics. This programme, Mathematics and Its Applications, is devoted to new emerging (sub)disciplines and to such (new) interrelations as exempla gratia:

- a central concept which plays an important role in several different mathematical and/or scientific specialized areas;
- new applications of the results and ideas from one area of scientific endeavour into another;
- influences which the results, problems and concepts of one field of enquiry have and have had on the development of another.

The Mathematics and Its Applications programme tries to make available a careful selection of books which fit the philosophy outlined above. With such books, which are stimulating rather than definitive, intriguing rather than encyclopaedic, we hope to contribute something towards better communication among the practitioners in diversified fields.

There are some seemingly quite specialized bits of theory in mathematics which are so central in so many different and widely-separated parts of mathematics and which relate less intimately to so many yet other bits and pieces, that one despairs - rightly - of ever being able to write down a really good classification scheme. One of these is the Riemann boundary value problem, also variously known as the Riemann-Hilbert problem, the (potential) barrier problem, and various other less-often used names. In one of its simpler matrix multiplicative forms it looks as follows. Given a smooth non-self-intersecting contour Γ on the Riemann sphere with interior region C_- and exterior region C_+ and given an invertible matrix-valued function $g(\lambda)$ on Γ_1 find nonsingular matrix-valued functions X_+ and X_- analytic in C_+ and C_-, respectively, such that their boundary values on Γ exist and such that $X_+ = X_-g$ on Γ. There are many generalizations: allowing (designated) plus and zeros in C_+ and C_-, more general contours, more (complex) variables instead of one, solutions in terms of special classes of functions such as almost-periodic or automorphic, and finally the same problem on surfaces of higher genus instead of the Riemann sphere of genus zero. The last generalization is the subject of this book.

The Riemann-Hilbert boundary value problem sounds quite specialized and thus it may come as a surprise that the problem and its applications generate a steady 45 papers a year. Still more surprising is the astonishing variety of fields to which it is narrowly related. Classically, the problem ties up narrowly with the theory of algebraic functions, the Jacobi inversion problem, the Abel theorem, the Riemann-Roch theorem, theta functions and the like. The best known approach to solving this problem is usually done in terms of singular integral equations, whence a whole stew of relations with that topic and integral transforms and also specially with Wiener-Hopf equations and factorizations of operator-valued functions. The last topic, factorizations, is something like a Bruhat decomposition of the infinite-dimensional loop groups. A special case of the Riemann-Hilbert boundary-value problem is the Riemann monodromy problem which asks for certain meromorphic multivalued functions with prescribed monodromy behaviour. The more dimensional generalizations of the problem involve holonomic systems, D-modules, perverse sheafs, ... (Riemann-Hilbert correspondence). There are applications of the Riemann-Hilbert problem to elasticity (and cracks), to electromagnetism (bonding to apertures, gratings), to Rayleigh scattering, to numerical methods in conformal mapping and for the determination of zeros, to optimal control and boundary control problems, to the heat conduction, to streamlines of flows, to the bending of plates, to pluriharmonic functions and to many more topics.

And last but not least the most elegant and promising (in my opinion) approach to the understanding of integrable systems and solitons, the Zakharov-Shabat dressing method, is totally based on the Riemann-Hilbert problem. The same idea ('dressing') occurs in the theory of the (axially symmetric) Einstein equations, where one speaks in this context of Hauser-Ernst and Kinnersley-Chitre transformations. It is for this last most important set of applications to soliton theory that the generalization to surfaces of higher genus is absolutely necessary. A number of completely integrable systems simply 'belongs' (in this sense) to surfaces of higher genus; this is far from the only reason for being interested in this version of the problem but it is a main one and these applications indeed form the concluding chapter of the book. The author has made fundamental contributions in this area starting in the 1950's and now has put everything together in this first monograph on the Riemann-Hilbert problem on Riemann surfaces and their applications, thus providing a convenient, coherent and compact starting volume for further research in this fascinating topic.

The unreasonable effectiveness of mathematics in science ...

Eugene Wigner

Well, if you know of a better 'ole, go to it.

Bruce Bairnsfather

What is now proved was once only imagined.

William Blake

As long as algebra and geometry proceeded along separate paths, their advance was slow and their applications limited.

But when these sciences joined company they drew from each other fresh vitality and thenceforward marched on at a rapid pace towards perfection.

Joseph Louis Lagrange.

Bussum, September 1987 Michiel Hazewinkel

CONTENTS

Preface xi

Chapter 1. The Riemann Boundary Problem on Closed Riemann Surfaces 1

§1. Riemann Surfaces 1
§2. Functions and Differential Forms. Abelian Integrals and Differentials 5
§3. Riemann Bilinear Relations. The Riemann–Roch Theorem 11
  A. Bilinear Relations 11
  B. A Differential Order 16
  C. The Riemann–Roch Theorem 17
§4. Cauchy-type Integrals 19
§5. The Riemann Problem. Number of Solutions 26
§6. Inversion of Abelian Integrals and Abel's Theorem. Solvability of the Riemann Problem 35
  A. Abel's Theorem 35
  B. Inversion of Abelian Integrals. The Boundary Problem Solvability 38
  C. Jacobi Variety 41
§7. Riemann Theta-Functions. Solvability of the Riemann Boundary Problem 42
  A. Zeros of the Riemann Theta-Function 42
  B. The Problem of Inversion of Abelian Integrals 44
  C. Divisor Classes 46
  D. The Solvability of the Riemann Problem 47
§8. Explicit Formulae for Solutions of the Riemann Problem 48

Chapter 2. Complex Vector Bundles over Compact Riemann Surfaces 54

§9. De Rham and Dolbeault Theorems 54
§10. Divisors. Complex Vector Bundles. Serre and Riemann Theorems 60
  A. Divisors and Complex Line Bundles 60
  B. The Serre Duality Theorem 62
  C. The Riemann Theorem 63
§11. The Riemann–Roch Theorem. The Riemann Problem 66
  A. The Riemann–Roch Theorem 66

B. Some Corollaries 67
C. The Riemann Boundary Problem 69
§12. The Second Cousin Problem. Solvability of the Riemann Problem 71
A. Characteristic Classes. Abel's Theorem 71
B. The Second Cousin Problem 73
C. Classification of Complex Line Bundles 77
D. Solvability of the Riemann Problem, $\kappa = 0$ 78
E. Solvability of the Riemann Problem, $0 < \kappa < g$ 80
F. The Nonhomogeneous Riemann Problem 82

Chapter 3. The Riemann Boundary Problem for Vectors on Compact Riemann Surfaces 83

§13. The Riemann Boundary Problem for Vector Functions 83
A. The Riemann Problem and Complex Vector Bundles 83
B. The General Solution of the Riemann Problem 88
C. The Conjugate Problem. The Riemann–Roch Theorem for Vector Bundles 91

Chapter 4. The Riemann Boundary Problem on Open Riemann Surfaces 94

§14. Open Riemann Surfaces 94
A. Finite Surfaces 94
B. Triviality of Cohomologies on Open Riemann Surfaces 96
C. The Riemann Bilinear Relations 99
D. The Hodge–Royden Theorem 102
§15. D-Cohomologies 102
A. D-Cohomology Groups. The Singular Group 102
B. Serre Duality 105
§16. D-Divisors. The Second Cousin Problem 106
A. Divisor Degree 106
B. Infinite Divisors 108
C. S-Divisors 111
D. The Second Cousin Problem 113
§17. The Riemann Problem. Solvability 116
A. The Problem Statement. The Bundle B_G 116
B. The Existence of a Solution 118
C. The Cauchy Index. The Solvability Conditions 120
D. The Case $\kappa = 0$. S-Problems 123
§18. The Solving of the Riemann Problem in the Explicit Form 127
A. Cauchy-type Integrals 127
B. Construction of a Solution 129

Chapter 5. Generalized Analytic Functions 132

§19. Bers–Vekua Integral Representations 132
A. Generalized Analytic Functions on a Plane 132
B. Generalized Analytic Functions on a Compact Riemann Surface. Basic Definitions 134

C. The First Bers–Vekua Equation 136
D. Equation $\bar{\partial} u = Au$ 137
§20. The Riemann–Roch Theorem 139
A. Generalized Constants 139
B. The Riemann–Roch Theorem 142
§21. Nonlinear Aspects of the Generalized Analytic Function Theory 146
A. Multiplicative Multivalued Solution. Existence 146
B. Multiplicative Constants. Uniqueness 147
C. Abel's Theorem 150

Chapter 6. Integrable Systems 151

§22. The Schrödinger Equation 151
A. Fast-Decreasing Potentials 151
B. Reflection Finite-Zone Potentials 158
§23. The Landau–Lifschitz Equation 176
A. Fast-Decreasing Potentials 167
B. Reflection Finite-Zone Potentials 171
§24. Riemann–Hilbert and Related Problems 172
A. D-Bar Problem 172
B. The Dressing Method 174
C. The Riemann–Hilbert Problem 175

Appendix 1 Hyperelliptic Surfaces 179
Appendix 2 The Matrix Riemann Problem on the Plane 181
Appendix 3 One Approximate Method of Solving the Matrix Riemann Problem 183
Appendix 4 The Riemann–Hilbert Boundary Problem 186

Notations 188

References 191

Subject Index 197

PREFACE

Let L be a closed contour separating the complex plane into the domains T^+ and T^-, and $G(t)$ be a matrix (or a function) defined on L. The Riemann boundary problem on the plane is to determine analytic matrices (functions) $F^{\pm}(z)$ in the domains $T^{\pm}$, respectively, satisfying the boundary condition

$$F^+(t) = G(t)F^-(t), \qquad t \in L.$$

This problem appears in different areas of mathematics and physics with a striking constancy. Singular integral equations, Wiener–Hopf operators, boundary properties of analytic functions. operator rings, elliptic systems—all of these areas are closely related to the Riemann boundary problem. This problem was formulated by B. Riemann and was studied by D. Hilbert, C. Hazeman, J. Plemelj, N. I. Muskhelishvili, F. D. Gahov, M. G. Krein, I. C. Gohberg, A. Grothendieck, B. V. Bojarskii, and many others. The Riemann problem has a wide range of physical applications, such as in contact problems of elasticity theory, dispersion relations in quantum mechanics, flow problems in hydrodynamics, diffraction theory, and so on. Recently, V. E. Zakharov, A. B. Shabat, and A. V. Mikhailov reduced the inverse scattering problem for the one-dimensional Schrödinger equation and for some other integrable systems to the Riemann boundary problem and turned it into a fundamental tool of the Hamiltonian system and soliton theory.

The study of the Riemann problem on Riemann surfaces was begun by W. Koppelman and the author at the end of the 1950s and was continued by L. I. Chibricova, R. N. Abdulaev, E. I. Zverovich, and others. This established the relations between the Riemann boundary problem and singular integral equations and the basic facts of algebraic function theory. In 1957 A. Grothendieck investigated connections between the Riemann problem on the plane and complex vector bundles. In 1962 H. Röhrl obtained analogous results for the Riemann problem and complex vector bundles over Riemann surfaces. New physical applications (for example, the inverse spectral problem for a Schrödinger equation with periodic potential, and some problems of field theory) have also appeared. Several of these relations can be formulated in the classical terms of Riemann surface theory (the Riemann–Roch and Abel theorems, the Abelian integral inversion, theta-functions, Jacobi varieties), others demand more modern technical tools, such as cohomologies with coefficients in sheaves.

Thus, the Riemann problem on Riemann surfaces became the centre of attention for many mathematicians and physicists. The present book is intended both for the specialists and for the first acquaintance with the subject. A preliminary knowledge of Riemann surfaces, algebraic topology, and scattering theory is not assumed. We omit many details which may be found in traditional textbooks, thus providing a more concise account.

The author attempts to give a very direct account of Riemann surfaces. When the central facts of algebraic function theory are equivalent to several aspects of the Riemann problem, the account of compact Riemann surface theory and the scalar Riemann problem are treated in parallel. Such an approach is economical and shows the essence of the matter. The first part of the book (Chapters 1 and 2) is dedicated to these problems. Chapter 1 contains the more classical aspects of Riemann surfaces: Abelian integrals, Cauchy kernels, the Riemann–Roch and Abel theorems, the Jacobi inversion problem, the Riemann theta-function, and corresponding facts for a Riemann problem (an index, solvability, explicit formulae).

Chapter 2 is devoted to the relations between the Riemann boundary problem and complex line bundles. Here we consider sheaves of germs of differential forms, divisors and corresponding line bundles, and cohomology groups. Simultaneously, the line bundle corresponding to the Riemann boundary problem and the sheaf of germs of its solutions are studied.

In Chapter 3 we shall look at the matrix Riemann boundary problem and its connections with complex vector bundles. This theory is far from complete. We shall study one of its aspects: partial indices, the structure of the general solution, and related problems. A number of very important problems (for example, Mumford stable bundles) are omitted, since their relations with the Riemann boundary problem are still vague.

Open surfaces of infinite genus are studied in Chapter 4. In this case the Cauchy–Riemann operator has an infinite index. This leads to specific problems; for example, solutions of the Riemann problem with an infinite number of zeros and poles. The standard cohomological methods of Chapter 2 are not applicable here, since all cohomology groups are equal to zero in this case. We combine this approach with the classical Ahlfors–Nevalinna method using a finite Dirichlet integral. This provides an investigation of the Riemann problem and a related area, including some exotic objects. For example, we define the degree of an infinite divisor (a divisor is a symbol describing zeros and poles of a function; the divisor degree is the difference of the sums of order of all zeros and poles) and consider infinite divisors of both finite and infinite degrees. We shall study Jacobi varieties of infinite dimension and such objects as the singular group having no analogs in the compact case.

Elliptic systems of the first order with two unknown functions are analogous to the Cauchy–Riemann system in many respects. The theory of generalized analytic (pseudoanalytic) functions of Bers–Vekua on the plane is widely known. In the case of Riemann surfaces, many new aspects have appeared. Multivalued generalized analytic functions can be different. The functions of the first type get an additional increment encircling a cyclic section of the surface (a closed path which is nonhomological to zero). Such functions are called integrals. Other functions get a factor in this situation. In the analytic case, these types of multivalued functions are related by the logarithm operation. For generalized analytic functions, the role of the logarithm is played by some nonlinear integral operator which is studied here. The index of the differential equation system is also calculated (the Riemann–Roch theorem).

Chapter 6 plays a special role in the book. It is dedicated to physical applications of the Riemann boundary problem on Riemann surfaces; its subject is completely integrable systems. In §22 we consider the inverse problem for the one-dimensional Schrödinger equation with a time-independent spectrum. In this case a potential satisfies the famous Korteweg–de Vries equation. For a fast-decreasing potential, the continuous spectrum is $(0, \infty)$, and the inverse scattering problem is reduced to the Riemann problem on the plane. If the continuous spectrum consists of $n(n > 1)$ zones, the inverse spectrum problem is reduced to the matrix Riemann problem on a hyperelliptic surface of the genus $n - 1$. In particular, if $G \equiv 1$, we get soliton $(n = 1)$ or finite-zone $(n > 1)$ reflectionless solutions. The Riemann problem method describes the corresponding Hamiltonian system completely.

The analogous situation for the Landau–Lifschitz equation, describing nonlinear waves in ferromagnets and some field theory models, is studied in §23. Finally, in §24, we consider the Riemann–Hilbert problem of analytic theory of differential equations and related problems.

The study of the Riemann boundary problem on Riemann surfaces is due to the initiative of L. I. Volkoviskii who foresaw many results and relations. The author repeatedly discussed this problem also with R. N. Abdulaev and S. Y. Gusman. The physical applications of the Riemann problem are based on ideas of S. P. Novikov and V. E. Zakharov. Numerous fruitful discussions with S. Y. Alber, V. P. Gurariǐ, A. R. Its, A. V. Mikhailov, V. I. Mačaev, V. B. Matveev, and V. E. Zakharov, and their constant attention and support were extremely important for the author.

The author would like to express his sincere gratitude to all these persons. The author would also like to thank the editors at Reidel Publishing and Rosenlaui Publishing Services for their help in producing this book.

CHAPTER 1

THE RIEMANN BOUNDARY PROBLEM ON CLOSED RIEMANN SURFACES

§1 Riemann Surfaces

A Riemann surface is a two-dimensional manifold having a complex structure. We now define these notions. A two-dimensional manifold M is a Haussdorf topological space on which every point $p \in M$ has a neighbourhood U_p homeomorphic to the unit disk $|z| < 1$ of the complex z-plane. The function $z(q)$ that homeomorphically maps U_p on $|z| < 1$ is called a local coordinate (local parameter), and U_p is a coordinate neighbourhood. Choose a covering of the surface M by coordinate neighbourhoods possessing the following property. Let the intersection of two coordinate neighbourhoods U and U' be nonempty. Then for every two local coordinates z and z' in $U \cap U'$, the correspondences $z = z(z')$ and $z' = z'(z)$ are defined. We demand that these functions (called relationships of neighbourhoods) should be holomorphic. A chosen class of local coordinates $\{U, z(p)\}$, called the atlas, determines the complex structure of the manifold M. We can choose other systems of local coordinates $\{V, w(p)\}$ such that all functions $z = z(w)$ are holomorphic. Such an atlas is equivalent to the first one and determines the same complex structure.

On a given manifold, different complex structures can be determined. For instance, two annuli, $1 < |z| < R, 1 < |w| < R^2, R \geqslant 1$, of the complex planes are topologically equivalent and are related by the (nonconformal) homeomorphism $w = |z|z$. Of course, the natural complex structures of each of the annuli generated by the embedding in the corresponding complex plane are different.

The most natural way leading to Riemann surfaces is an investigation of multivalued functions and analytic continuation.

Let the power series

$$f_U(z) = \sum_{k=0}^{\infty} c_k (z - z_0)^k \tag{1.1}$$

converge in the disk U of the complex plane. The pair $\{U, f_U\}$ is known as the functional element of Weierstrass. Now let L be some closed curve $z_0 \in L$. Continuing the functional element along the curve L, we obtain a new functional element $\{V, f_V(z)\}$

$$f_V(z) = \sum_{k=0}^{\infty} c'_k (z - z_0)^k,$$

which, in general, is different from (1.1). Considering the set of pairs $\{z, \{U, f_U\}\}$, we can introduce on this set the complex structure corresponding to a multi-sheated surface over the z-plane (see, for example, Nevalinna [19]).In the most interesting case, when the surface is determined by the equation

$$P(z, w) = 0, \tag{1.2}$$

$P(z, w)$ is a polynomial of two variables. Here we note that this surface is compact. Conversely, any compact Riemann surface is determined by the algebraic equation of the type (1.2). For example, the equation

$$w^2 = (z - z_1)(z - z_2)(z - z_3)(z - z_4) \tag{1.3}$$

determines the two-sheated surface over the z-plane obtained by splicing of two copies of the z-plane cut along the lines connecting points z_1, z_2 and z_3, z_4. As one can see in Figure 1, this surface is topologically equivalent to a torus.

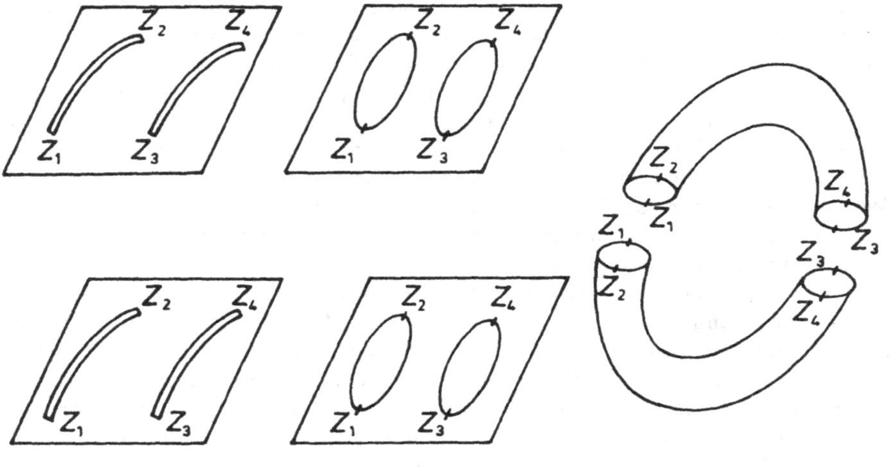

Figure 1

Now we summarize some facts about the surface topology, which can be found in the text for any standard course (see, for example, Springer [21]).

A closed (compact) Riemann surface is homeomorphic to the sphere with g handles. The number g is called a genus of the surface. For $g = 0$, we have a Riemann sphere, and for $g = 1$, a torus. A characteristic property of all surfaces, for $g > 0$, is the existence of cyclic sections, i.e., closed curves not dividing the surface into two parts (see Figure 2). For each handle, there exist two such oriented sections (a parallel and a meridian of a torus). These sections will be numbered such that every even cycle crosses the corresponding odd one from right to left (in topological terms this means that the intersection index $I(k_{2j-1}, k_{2j}) = 1$) and intersects no other cycles.

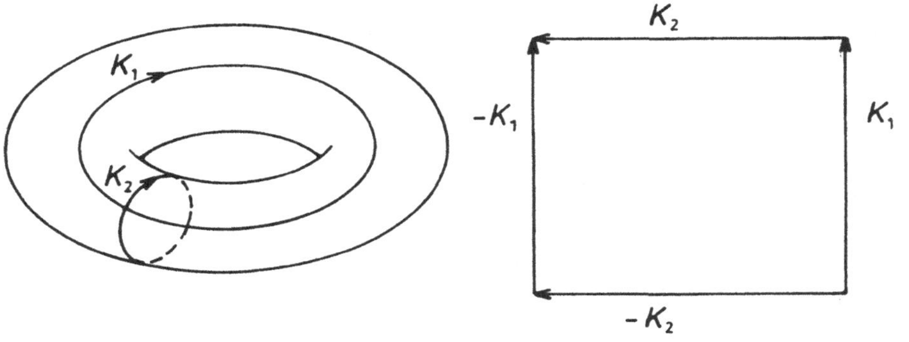

Figure 2

Deform $2g$ of these cycles so that they intersect each other at one point, and cut the surface along these cycles (Figures 2 and 3). We obtain a $4g$-sided polygon with pairs of sides oriented to meet each other. Below, we shall fix these orientations and shall distinguish the sides of the polygon corresponding to different sides of the cut by the signs $\pm$.

It can be shown that the cycles $k_1, \ldots, k_{2g}$ form a basis for the one-dimensional homology group $H_1(M)$ (the Betti group). The reader may think of the elements of the homology group as linear sums of the type $\sum_{j=1}^{2g} c_j k_j$, where c_j are elements of the base ring (for example, real or complex numbers). This basis is called canonical. In the following, we shall fix this basis. The surface obtained from the surface M by cutting along the canonical basis is denoted by $\widehat{M}$. The representation of the surface M by the polygon involves the possibility of its triangulation. A triangulation of a surface is the division of the surface into a denumerable (finite for compact surfaces) set of closed triangles satisfying the following conditions:

a) the triangles have no common interior points,
b) two triangles can possess a common side or a common vertex,
c) every vertex belongs to a finite number of triangles.

In the following we suppose that the triangle boundaries are oriented such that the interior of the triangle is at the left of a boundary cycle. A prearranged

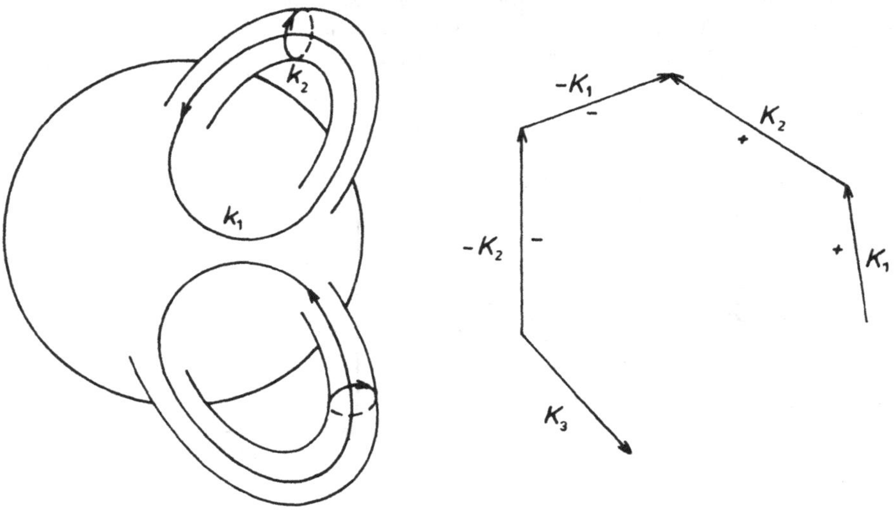

Figure 3

choice of orientations of triangles (when common boundary sides of triangles are passed in the opposite directions) is possible for the class of so-called orientable surfaces. Riemann surfaces belong to this class.

Let some triangulation be fixed, and let a^0, a^1, a^2 be numbers of 0-, 1-, and 2-simplexes (vertexes, sides, and triangles) forming this division. The sum

$$\chi = -a^2 + a^1 - a^0 \tag{1.4}$$

is called the Euler characteristic of the surface. Its value depends only on the surface genus g and is equal to

$$\chi = 2g - 2. \tag{1.5}$$

§2 Functions and Differential Forms. Abelian Integrals and Differentials

Let M be a closed Riemann surface and $D \subset M$ be a domain belonging to M (possibly, $D = M$). An analytic structure allows the consideration of functions possessing any smoothness including analytic functions, since such functions conserve their smoothness if one makes local coordinate changes in the fixed atlas.[1]

As with functions, we shall consider differential forms of the type

$$\omega = a(z)dz(p) + b(z)\overline{dz(p)}$$

invariant with respect to local coordinate changes $z(p)$ on $z^*(p)$. The differential of the function $f(p)$,

$$df = \frac{\partial f(p)}{\partial z(p)} dz(p) + \frac{\partial f(p)}{\partial \overline{z(p)}} \overline{dz(p)},$$

can be considered as an example of such a form. Below we shall use the notations dp and $d\bar{p}$ or dz and $d\bar{z}$ in place of $dz(p)$ and $\overline{dz(p)}$. It is clear that if we make a local coordinate change, the coefficients of a differential form must be changed by the rule

$$a(z^*) = a(z)\frac{dz(p)}{dz^*(p)}, \qquad b(z^*) = b(z)\frac{\overline{dz(p)}}{\overline{dz^*(p)}}.$$

First we consider differential forms of the class C^1. If s some sufficiently smooth curve in the domain D, the integral $\int_s \omega$ along s is defined. This integral is independent of a continuous deformation of the integration contour if and only if the condition

$$\frac{\partial a}{\partial \bar{z}} - \frac{\partial b}{\partial z} = 0 \tag{2.1}$$

is valid. Differentials of the class C^1 satisfying the condition (2.1) are called closed differentials. If ω is closed, we can consider the primitive

$$\Omega(p) = \int_{p_0}^{p} \omega$$

with a fixed low limit. The value $\Omega(p)$ is a function of the point $p, d\Omega = \omega$. However, in general, the function $\Omega(p)$ is multivalued in the domain D, since the function $\Omega(p)$ increases by $\oint_k \omega$ when the argument goes along a cycle k nonhomological to zero. This value is called the period of ω along k. Values $\Omega(p)$, such that $d\Omega$ are closed differentials, are called integrals. We often denote closed differentials by $d\Omega$, assuming that Ω is the corresponding integral. If all

[1]Below we permit only such changes of coordinates.

periods of the differential $\omega = d\Omega$ are equal to zero, the integral $\Omega(p)$ is single-valued in the domain D and is a well-defined function in D. The differential form ω is called exact in this case.

For every form $\omega = adz + bd\bar{z}$, we define the form

$$*\omega = \frac{1}{2i}(\bar{b}dz - \bar{a}d\bar{z}). \tag{2.2}$$

It is clear that $*\omega$ is an invariant differential form, and $**\omega = -\frac{1}{4}\omega$. If the form $*\omega$ is closed, the form ω is called coclosed (correspondingly, coexact). If the form $\omega \in C^1$ is closed and coclosed simultaneously, then by (2.1), we have

$$\frac{\partial a}{\partial z} = 0, \quad \frac{\partial b}{\partial \bar{z}} = 0. \tag{2.3}$$

Such differentials are called harmonic. The corresponding integrals, as is easily seen, are harmonic functions of local coordinates.

Having two differentials ω_1 and ω_2, we can construct a quadratic differential $\omega_1 \wedge \omega_2$, called an exterior product of ω_1 and ω_2. The exterior product is determined by two conditions: linearity on both multipliers and anticommutativity, $\omega_1 \wedge \omega_2 = -\omega_2 \wedge \omega_1$. Denoting $z = x + iy$, we identify the differential $\frac{1}{2i}d\bar{z} \wedge dz = dx \wedge dy$ with the area element. Thus we have

$$\omega_1 \wedge \omega_2 = (a_2 b_1 - a_1 b_2)d\bar{z} \wedge dz,$$

since the anticommutativity involves $dz \wedge dz = d\bar{z} \wedge d\bar{z} = 0$. We also define the operation of exterior differentiation leading to quadratic differentials by

$$d\omega = d(adz + bd\bar{z}) \overset{\text{def.}}{=} da \wedge dz + db \wedge d\bar{z} = \left(\frac{\partial a}{\partial \bar{z}} - \frac{\partial b}{\partial z}\right) d\bar{z} \wedge dz. \tag{2.4}$$

Therefore, closed differentials may be defined by the condition $d\omega = 0$.

If $T \subset D$ is a domain and ∂T is its boundary, Green's formula is valid:

$$\int_T d\omega = \int_{\partial T} \omega. \tag{2.5}$$

Now consider the Hilbert space $H(D)$ of differentials whose coefficients are square-integrable over D (it should be recalled that D can coincide with M). The scalar product is defined by the formula

$$(\omega_1, \omega_2) = \int_D \omega_1 \wedge *\omega_2 = \int_D (a_1\bar{a}_2 + b_1\bar{b}_2)dx \wedge dy. \tag{2.6}$$

The closure of the linear set of closed differentials of the class C^1 in the metric of the space $H(D)$ is denoted by $\Gamma_c(D)$. Elements of the space $\Gamma_c(D)$ are also called closed differentials.

For a smooth curve $s \subset D$, we consider the value $\int_s \omega, \omega \in H(D)$. It is clear that for $\omega \in \Gamma_c(D)$, the primitive

$$\Omega(p) = \int_{p_0}^{p} \omega$$

in general, is multivalued in D. For every cycle k noncohomological to zero, we define the linear functional on $\Gamma_c(D)$,

$$k[\omega] = \int_k \omega, \qquad \omega \in \Gamma_c(D). \tag{2.7}$$

The value $k[\omega]$ is called the period of the differential ω (or of a corresponding integral) along the cycle k. A closed differential whose periods are equal to zero is called *exact*. The space of the exact differentials in D is denoted by $\Gamma_e(D)$. The closure of the linear set of coclosed differentials of the class C^1 is called the space of coclosed differentials and is denoted by $\Gamma_c^*(D)$. The space of coexact differentials is denoted by $\Gamma_e^*(D)$. The famous Weyl lemma asserts that the intersection of these spaces $\Gamma_c(D) \cap \Gamma_c^*(D) = \Gamma_h(D)$ is the space of harmonic differentials in D. Here it is necessary to verify that any differential belonging to this intersection is smooth and consequently has analytic coefficients because of (2.3). The reader can find the proof of this lemma in standard texts on Riemann surfaces (see, for example, [8] and [21]). On a closed surface, the following theorem for harmonic differentials is valid.

THEOREM 2.1 (LIOUVILLE). *A harmonic function on a closed Riemann surface is a constant. The intersection $\Gamma_e(M) \cap \Gamma_h(M)$ is zero.*

In fact, a harmonic function on M takes its extreme values only at boundary points of the surface. But such points are absent in the case of the closed surface. If $\omega \in \Gamma_e(M) \cap \Gamma_h(M)$, then $\omega = dh$, where h is a harmonic function, and $\omega = 0$.

THEOREM 2.2 (HODGE). *The direct decomposition*

$$H(M) = \Gamma_e(M) \oplus \Gamma_e^*(M) \oplus \Gamma_h(M) \tag{2.8}$$

is valid for a compact surface M.

The orthogonality of these spaces is verified directly. For example, let $dh \in \Gamma_e(M)$ and $\omega \in \Gamma_h(M)$. Then

$$(dh, \omega) = \int_M dh \wedge *\omega = -\int_M h \wedge d(*\omega) = 0.$$

The intersection of these spaces is zero. In fact, $\Gamma_e(M) \cap \Gamma_e^*(M)$ contains only exact harmonic differentials. All these differentials are equal to zero, as was shown above. If $\omega \in \Gamma_e^*(M) \cap \Gamma_h(M)$, then $*\omega$ is an exact harmonic differential, and $\omega = 0$. From (2.7) the relation

$$\Gamma_c(M) = \Gamma_e(M) \oplus \Gamma_h(M) \tag{2.8'}$$

follows.

Let $d\Omega$ be a closed differential having preset periods. We obtain the representation

$$d\Omega = df + ds,$$

where f is a single-valued function and ds is a harmonic differential. Therefore, we obtain the following theorem.

THEOREM 2.2′. *On a closed Riemann surface M, there exists a harmonic differential having prescribed periods along cycles of a homological basis.*

A harmonic differential of the type

$$\omega = a dz, \qquad \partial a / \partial \bar{z} = 0$$

is called *analytic*. An analytic differential is represented as the sum

$$\omega = \frac{1}{2}(a dz + a d\bar{z}) + \frac{1}{2}(a dz - \bar{a} d\bar{z})$$

of two harmonic differentials (real and imaginary parts).

In parallel with regular analytic differentials, we shall study meromorphic differentials possessing poles. Such a differential has an expansion in the neighbourhood of a pole q of n-th order,

$$\omega(p) = \sum_{k=1}^{n} \frac{c_{-k} dz}{[z(p) - z(q)]^k} + \text{analytic part.} \tag{2.9}$$

Here $z(p)$ is a local coordinate. By changing a local coordinate, we change the coefficients c_{-k} of the principal part, except the residue c_{-1} determined by the relation

$$c_{-1} = \frac{1}{2\pi i} \int_{l_q} \omega, \tag{2.10}$$

where l_q is a closed contour around q.

Note. The formula (2.10) conserves its meaning when ω is a closed differential. The structure of a singularity of such a differential can be very intricate. We call *singularities* only such points in which the local Dirichlet integral (2.6) diverges.

Differentials that are analytic on the closed surface, except for a finite number of points (where one may have poles), are called *Abelian differentials*, and the corresponding integrals are called *Abelian integrals*. Abelian differentials that are analytic everywhere on the surface are called *Abelian differentials of the first kind*. Abelian differentials on the surface having a finite number of poles with zero residues are called *Abelian differentials of the second kind*. Abelian differentials having poles with nonzero residues are called *Abelian differentials of the third kind*.

Note that the sum of the residues of an Abelian differential is equal to zero. This results from the following theorem.

THEOREM 2.3. *The sum of the residues of a closed differential with a finite number of singularities is equal to zero.*

In fact, triangulate the surface and denote the 2-simplexes of the triangulation by U_j. We obtain the obvious equation

$$\sum_j \int_{\partial U_j} \omega = 0,$$

since every side of the 2-simplex is passed twice in opposite directions.

THEOREM 2.4 (RIEMANN). *The real dimension of the linear space of Abelian differentials of the first kind is equal to $2g$, where g is a genus of the surface M.*

It is almost obvious that this number does not exceed $2g$. In the opposite case, it is possible to construct the Abelian differential of the first kind with $2g$ purely imaginary periods. The real part of such a differential is an exact harmonic differential. According to Theorem 1.1, this differential is equal to zero.

In order to complete our proof, it is necessary to construct $2g$ linearly independent Abelian differentials of the first kind.

Let $k_1, \ldots, k_{2g}$ be a canonical homological basic of M. Due to the Hodge theorem (Theorem 2.2) there exist real harmonic differentials η_j with periods

$$\int_{k_s} \eta_j = \delta_{ij}, \qquad s, j = 1, \ldots, 2g. \tag{2.9}$$

By using harmonic conjugates to obtain analytic differentials, we obtain the Abelian differentials of the first kind,

$$dW_j = \eta_j + i(*\eta_j), \qquad j = 1, \ldots, 2g. \tag{2.10}$$

This set forms the basis for the $2g$-dimensional space of Abelian differentials of the first kind.

Each basis of this space determines a $2g \times 2g$-matrix of periods. In our case it has the form

$$\left(\int_{k_s} dW_j\right), \quad \operatorname{Re}\int_{k_s} dW_j = \delta_{sj}, \qquad s, j = 1, \ldots, 2g. \tag{2.11}$$

By a linear transformation, the matrix (2.11) can be reduced to the form

$$\begin{pmatrix} 1 & 0 & \ldots & 0 & c_{1,g+1} & \ldots & c_{1,2g} \\ c_{21} & \ldots & \ldots & c_{2g} & c_{2,g+1} & \ldots & c_{2,2g} \\ 0 & 1 & \ldots & 0 & c_{3,g+1} & \ldots & c_{3,2g} \\ \vdots & \vdots & \vdots & \vdots & \vdots & \vdots & \vdots \\ 0 & 0 & \ldots & 1 & c_{2g-1,g+1} & \ldots & c_{2g-1,2g} \\ c_{2g,1} & \ldots & \ldots & c_{2g,g} & c_{2g,g+1} & \ldots & c_{2g,2g} \end{pmatrix}, \tag{2.12}$$

where

$$c_{ij} \equiv \int\limits_{k_s} d\tilde{\omega}_i, \quad \begin{matrix} s = 2j-1 & j = 1, \dots g \\ s = 2(j-g) & j = g+1, \dots 2g, \quad i = 1, \dots g. \end{matrix}$$

Obviously, the corresponding basis of the differentials $d\tilde{\omega}_1, \dots, d\tilde{\omega}_{2g}$ has the following normalization of periods

$$\int\limits_{k_{2\mu-1}} d\tilde{\omega}_{2\nu-1} = \delta_{\mu\nu}, \qquad \mu, \nu = 1, \dots, g.$$

In the following section we shall show that the Abelian differentials of the first kind, whose periods along all cycles $k_{2j-1} (j = 1, \dots, g)$ are equal to zero, is zero. Therefore, denoting

$$d\tilde{\omega}_{2\nu-1} = d\omega_\nu, \qquad \nu = 1, \dots, g,$$

we obtain the complex basis of the space of Abelian differentials of the first kind with a normalization

$$\int\limits_{k_{2\mu-1}} d\omega_\nu = \delta_{\mu\nu}, \qquad \mu, \nu = 1, \dots, g. \tag{2.13}$$

Note also that the real basis determined by the normalization

$$\begin{aligned} &\operatorname{Im} \int\limits_{k_{2\mu}} d\theta_{2\nu-1} = -\delta_{\mu\nu}, \qquad \operatorname{Im} \int\limits_{k_{2\mu-1}} d\theta_{2\nu-1} = 0, \\ &\operatorname{Im} \int\limits_{k_{2\mu}} d\theta_{2\nu} = \delta_{\mu\nu}, \qquad \operatorname{Im} \int\limits_{k_{2\mu}} d\theta_{2\nu} = 0, \\ &\mu, \nu = 1, \dots, g. \end{aligned} \tag{2.13'}$$

We now introduce the normalized Abelian differentials and integrals of the second and third kinds. Let p be a point on the surface and $z(p)$ be a fixed local coordinate in the vicinity of this point. The normalized Abelian differentials of the second kind $dt^n_{p,z}(q)$ and $dT^n_{p,z}(q)$ are determined in the following manner. Some have a single pole of order $n+1$ at the point $q = p$, with a principal part

$$\text{princ. part } dt^n_{p,z} = \text{princ. part } dT^n_{p,z} = -\frac{n dz(q)}{[z(q) - z(p)^{n+1}} \tag{2.14}$$

and periods normalized by the relations

$$\begin{aligned} &\int\limits_{k_{2\mu-1}} dt^n_{p,z}(q) = 0, \qquad \mu = 1, \dots, g, \\ &\operatorname{Re} \int\limits_{k_j} dT^n_{p,z}(q) = 0, \qquad j = 1, \dots, 2g. \end{aligned} \tag{2.15}$$

For the case $n = 1$, we omit the upper index and the symbol of a local coordinate and write dt_p and dT_p.

Let p_0 and p be two arabitrary different points of the surface. The normalized Abelian differentials of the third kind, $d\omega_{p_0p}(q)$ and $d\Omega_{p_0p}(q)$, have poles with the residues -1 and $+1$ at the points p_0 and p, respectively, and periods satisfying the conditions

$$\int\limits_{k_{2\mu-1}} d\omega_{p_0p}(q) = 0, \qquad \mu = 1,\dots,g,$$

$$\operatorname{Re}\int\limits_{k_j} d\Omega_{p_0p}(q) = 0, \qquad j = 1,\dots,2g. \tag{2.16}$$

The existence of such differentials of the second and third kinds is proved in the theory of Riemann surfaces [22]. Usually, the existence of harmonic functions with corresponding singularities is proved, and then, using normalizations (2.15) and (2.16), adding some combinations of Abelian differentials of the first kind.

In §3 we shall show the existence of these differentials by other methods, using the Riemann–Roch theorem.

§3 Riemann Bilinear Relations. The Riemann–Roch Theorem

A. *Bilinear Relations*

Let df and dg be two closed differentials on the closed Riemann surface M having, in general, a finite number of singularities on M and satisfying the condition

$$df \wedge dg = 0. \tag{A}$$

Cutting the surface M along the cycles of a canonical homological basis, we obtain (see §1) the $4g$-sided polygon $\widehat{M}$ with the sides $k_1, k_2, -k_1, -k_2, k_3, \dots, -k_{2g}$. If all residues of df and dg are equal to zero, the integrals f and g are single-valued on $\widehat{M}$. On the identified sides of the polygon, the values of f and g are differentiated in values of periods along dual cycles. If the differentials df and dg have nonzero residues, then the integrals f and g have corresponding logarithmic singularities. In order to provide its single-valuedness, it is necessary to cut the polygon $\widehat{M}$ along some curves connecting the singular points of the integrals f and g. Denote the cut by s. Now consider the single-valued differential fdg in the cut polygon $\widehat{M}$. If the differential is closed, i.e.,

$$d(fdg) = df \wedge dg + f \wedge d^2 g = 0,$$

we have the relation

$$\int\limits_{\partial\widehat{M}+(s^+-s^-)} fdg = 2\pi i \sum_M \operatorname{res} fdg. \tag{3.1}$$

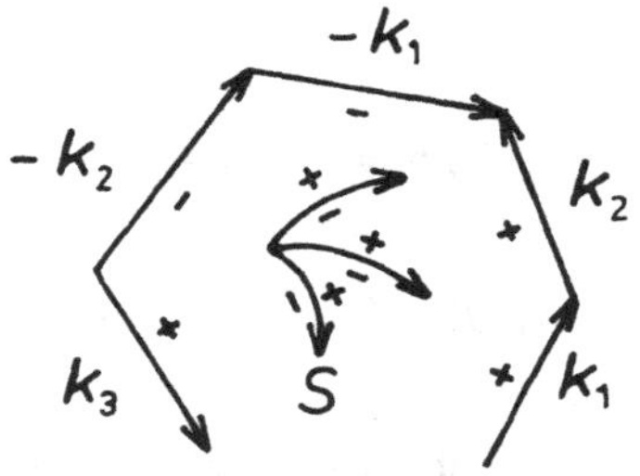

Figure 4

Here $s^{\pm}$ are sides of s. Let $(k)_\mu = k_{2\mu-1} + k_{2\mu} - k_{2\mu-1} - k_{2\mu}$. Then

(3.2)

$$\begin{aligned}\int_{\partial\widehat{M}} f\,dg &= \sum_{\mu=1}^{g} \int_{(k)_\mu} f\,dg \\ &= \sum_{\mu=1}^{g} \left\{ \int_{k_{2\mu-1}} \left[f\,dg - (f + \int_{k_{2\mu}} df)dg \right] + \int_{k_{2\mu}} \left[f\,dg - (f - \int_{k_{2\mu-1}} df)dg \right] \right\} \\ &= \sum_{\mu=1}^{g} \left[\int_{k_{2\mu-1}} df \int_{k_{2\mu}} dg - \int_{k_{2\mu}} df \int_{k_{2\mu-1}} dg \right] \stackrel{\text{def}}{=} L(f,g).\end{aligned}$$

Thus, we obtain the relation

$$L(f,g) + \int_s \Delta f\,dg = 2\pi i \sum_M \operatorname{res} f\,dg, \tag{3.3}$$

where Δf is the difference of values of f on both sides of the cut s. This formula is known as the first Riemann bilinear periods relation.

The second bilinear relation arises from Green's formula

$$\int_{\partial\widehat{M}} f\,dg = \int_{\widehat{M}} df \wedge dg. \tag{3.4}$$

This relation is valid for any $df, dg \in \Gamma_c$ From (3.2) it follows that

$$\int_{\widehat{M}} df \wedge dg = L(f,g). \tag{3.5}$$

In particular, for $g = \bar{f}$, we have

$$\frac{1}{2i}\int\limits_{\widehat{M}} d\bar{f} \wedge df = \sum_{\mu=1}^{g} \operatorname{Im}\left[\int\limits_{k_{2\mu-1}} d\bar{f} \int\limits_{k_{2\mu}} df\right] \tag{3.5'}$$

(the second Riemann periods relation).

From (3.5′) in particular, it follows that if all periods of an Abelian differential of the first kind are real or purely imaginary, then such a differential is zero.

Let df be an Abelian differential of the first kind, represented by

$$df = \sum_{\nu=1}^{g} c_\nu dw_\nu,$$

where c_ν are real numbers and $dw_\nu (\nu = 1, ..., g)$ is the basis (2.13). It follows from (3.6) that

$$\operatorname{Im}\sum_{\mu=1}^{g} \int\limits_{k_{2\mu-1}} d\bar{f} \int\limits_{k_{2\mu}} df \geqslant 0.$$

We obtain from (2.13) that

$$\operatorname{Im}\sum_{\mu=1}^{g} c_\mu \left(\sum_{\nu=1}^{g} c_\nu \int\limits_{k_{2\mu}} dw_\nu\right) \geqslant 0,$$

that is,

$$\operatorname{Im}\sum_{\mu\nu=1}^{g} c_\mu c_\nu \int\limits_{k_{2\mu}} dw_\nu \geqslant 0. \tag{3.6}$$

Therefore, the matrix

$$\left(\operatorname{Im}\int\limits_{k_{2\mu}} dw_\nu\right)$$

is positive-definite. This fact plays an important role in the theory of theta-functions (see §7).

The relation (3.3) enables us to calculate periods of normalized Abelian differentials.

Assuming $df = dw_\mu$, $dg = dw_\nu$ (see (2.13)), we obtain from (3.3) (cut s is absent, all residues are zero) that

$$\int\limits_{k_{2\mu}} dw_\nu = \int\limits_{k_{2\nu}} dw_\mu, \qquad \mu, \nu = 1, ..., g. \tag{3.7}$$

Note that (2.13) involves also the relations

$$\int_{k_{2\mu-1}} dw_\nu = \int_{k_{2\nu-1}} dw_\mu = \delta_{\mu\nu}, \qquad \mu,\nu = 1,\ldots,g. \tag{3.7'}$$

If we assume that $df = d\theta_{2\mu}, dg = d\theta_{2\nu}$ (see (2.13')) and calculate the imaginary part of $L(\theta_{2\mu}, \theta_{2\nu})$, we obtain

$$\int_{k_{2\mu}} d\theta_{2\nu} = \int_{k_{2\nu}} d\theta_{2\mu}, \qquad \mu,\nu = 1,\ldots,g. \tag{3.8}$$

In the same way, we conclude

$$\int_{k_{2\nu-1}} d\theta_{2\mu-1} = \int_{k_{2\mu-1}} d\theta_{2\nu-1}, \qquad \mu,\nu = 1,\ldots,g. \tag{3.9}$$

Assuming $df = d\theta_{2\mu-1}, dg = d\theta_{2\mu}$, we obtain

$$\int_{k_{2\nu}} d\theta_{2\mu-1} = \int_{k_{2\mu-1}} d\theta_{2\nu},\ \mu \neq \nu, \qquad \mu,\nu = 1,\ldots,g. \tag{3.10}$$

For $\mu = \nu$ we have

$$\operatorname{Re}\int_{k_{2\mu}} d\theta_{2\mu-1} = \operatorname{Re}\int_{k_{2\mu-1}} d\theta_{2\mu}, \qquad \mu = 1,\ldots,g.$$

Taking into account (2.13'), we conclude that

$$\int_{k_{2\mu}} d\theta_{2\mu-1} = \int_{k_{2\mu-1}} \overline{d\theta}_{2\mu}, \qquad \mu = 1,\ldots,g. \tag{3.11}$$

Assume $df = d\Omega_{q_1q_2}(p)$ (see (2.16)) and $dg = d\theta_j$ in the bilinear relation (3.3), and draw the cut s from point q_1 to point q_2, as in Figure 5. Then

$$\int_s \Delta f dg = \int_{q_1}^{q_2} f^+ dg + \int_{q_2}^{q_1} f^- dg,$$

where $f^\pm(p)$ are values of the integral $f(q)$ on the upper and lower sides of the cut, correspondingly. We have

$$\Delta f = f^+(p) - f^-(p) = \Delta\Omega_{q_1q_2}(p) = 2\pi i,$$

and

$$\int_s \Delta f dg = 2\pi i \int_{q_1}^{q_2} d\theta j.$$

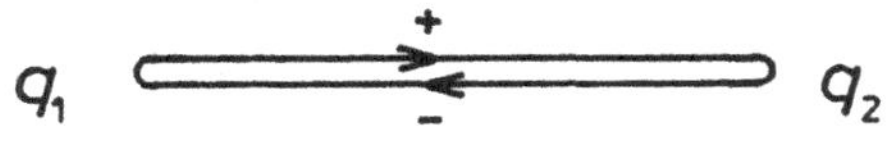

Figure 5

The residues of the differential $\Omega_{q_1q_2}(p)d\theta_j(p)$ are equal to zero,

$$\operatorname{Re} L\left(\Omega_{q_1q_2}, \theta_j\right) = -i \int_{k_j} d\Omega_{q_1q_2}.$$

Substituting these relations into (3.3), we obtain the relation

$$\int_{k_j} d\Omega_{q_1q_2}(p) = 2\pi i \operatorname{Im} \int_{q_1}^{q_2} d\theta_j, \qquad j = 1, \ldots, 2g. \tag{3.12}$$

Assuming $df = d\omega_{q_1q_2}$ and $dg = d\omega_\mu$, we obtain, in the analogous manner,

$$\int_{k_{2\mu}} d\omega_{q_1q_2}(p) = 2\pi i \int_{q_1}^{q_2} d\omega_\mu, \qquad \mu = 1, \ldots, g. \tag{3.13}$$

Assuming $df = dt_q^n(p)$ (see (2.15)) and $dg = dw_\mu$, we have

$$L\left(t_q^n, w_\mu\right) = -\int_{k_{2\mu}} dt_q^n(p),$$

$$\operatorname*{Res}_{p=q} t_q^n dw_\mu = \frac{1}{(n-1)!} \frac{d^n w_\mu(q)}{d(z(q))^n},$$

which involes the relations

$$\int_{k_{2\mu}} dt_q^n(p) = -\frac{2\pi i}{(n-1)!} \frac{d^n w_\mu(q)}{d(z(q))^n}, \qquad \mu = 1, \ldots, g. \tag{3.14}$$

If we assume that $df = dT_q^n(p)$ and $dg = d\theta_j$, then

$$\int_{k_j} dT_q^n(p) = -\frac{2\pi i}{(n-1)!} \operatorname{Im} \frac{d^n \theta_j(q)}{d(z(q))^n}, \qquad j = 1, \ldots, 2g. \tag{3.14'}$$

In conclusion, we assume that $df = d\omega_{q_1q_2}, dg = dt_p^n$. We have, as above,

$$\int_s \Delta f dg = 2\pi i \int_{q_1}^{q_2} dt_p, \qquad \operatorname{Res} f dg = -\frac{1}{(n-1)!} \frac{d^n \omega_{q_1q_2}(p)}{d(z(p))^n},$$

$$L(f, g) = 0.$$

We obtain the relation

$$\int_{q_1}^{q_2} dt_p^n = -\frac{1}{(n-1)!}\,\frac{d^n\omega_{q_1q_2}(p)}{d(z(p))^n}. \tag{3.15}$$

B. *A Differential Order*

An order $\operatorname{ord}(\omega)$ of the analytic differential ω is the difference between the sums of orders of all its zeros and poles. We shall show that

$$\operatorname{ord}(\omega) = 2g - 2.$$

First note that for the differential $d\ln f(p)$, where $f(p)$ is an analytic function, the theorem of residues is valid. Therefore, the sum of orders of zeros of $f(p)$ is equal to the sum of orders of its poles, i.e., $\operatorname{ord}(f) = 0$. It is clear that the function $f(p)$ takes any value the same number of times. Further, it is clear that orders of all differentials coincide since the ratio of any two differentials is a function on M. Therefore, it is sufficient to take an arbitrary analytic function $f(p)$ and to calculate $\operatorname{ord}(df)$.

Let the function $f(p)$ take every value n times. Denote orders of zeros and poles of the differential df by N_j and P_k. The corresponding points are branch points of the function $w = f(p)$ of orders N_j and $P_k - 2$ respectively,

$$\sum_j (N_j + 1) = \sum_k (P_k - 1) = n.$$

Consider a triangulation of the w-plane for which all branch points of the mapping $w = f(p)$ are vertices. We have the Euler relation for the numbers of simplexes

$$a^0 - a^1 + a^2 = q$$

(see (1.3), (1.4)). The considered triangulation determines the triangulation of an n-sheated surface over the w-plane which is an image of the surface M for the mapping $w = f(p)$. We have the Euler relation for numbers of simplices of this triangulation,

$$\alpha^0 - \alpha^1 + \alpha^2 = 2 - 2g, \qquad \alpha^1 = na^1, \qquad \alpha^2 = na^2.$$

The number of vertices less than na^0, since some are branch points. Thus we have

$$\alpha^0 = na^0 - \sum_j N_j - \sum_k (P_k - 2),$$

and

$$2 - 2g = n\left(\alpha^0 - \alpha^1 + \alpha^2\right) - \sum_j N_j$$

$$-\sum (P_k - 2) = 2n - \left(\sum_j N_j - \sum_k P_k\right) - 2\sum_k (P_k - 1) = -\operatorname{ord}(df).$$

This relation agrees with (3.16).

C. *The Riemann–Roch Theorem*

A divisor on the surface M is a finite sum $D = \sum c_k p_k$, where p_k are points of $M, p_k \in M$, and c_k are intergers. In a natural way we can introduce an addition of divisors generating the Abelian group $\operatorname{Div}(M)$. The divisor $D \geqslant 0$ if all $c_k \geqslant 0$. For every divisor D, we define the interger called the divisor degree, $\deg D = \sum c_k$.

Every analytic function $f(p)$ determines the divisor of its zeros and its poles, denoted by $(f) = \sum c_k p_k$, where p_k are zeros and poles of the function f, and $c_k = (\operatorname{ord} f)_{p_k}$ are orders of the function f at the points p_k. We see that the order of zeros are positive and the orders of poles are negative. It is obvious that $\deg(f) = 0$. If ω is an Abelian differential, then, according to (3.16), $\deg(\omega) = 2g - 2$.

We say that the function (or differential) is multiples of the divisor D, if $(f) \geqslant D$, i.e., $(f) - D \geqslant 0$.

Let D be some divisor. Consider the complex linear spaces $L(D)$ of analytic functions which are multiples of the divisor $-D$ and the complex space $H(D)$ of Abelian differentials which are multiples of the divisor D.

THEOREM 3.1 (RIEMANN–ROCH). *The relation*

$$\dim L(D) - \dim H(D) = \deg D - g + 1 \tag{3.17}$$

is valid.

Let

$$D = D' - D'', \qquad D' = \sum_{j=1}^{n} \alpha_j q_j \geqslant 0, \qquad D'' = \sum_{k=1}^{m} \beta_k p_k \geqslant 0.$$

Consider the space of Abelian integrals of the second kind which are multiples of the divisor $-D'$ having zero periods along the cycles $k_{2\mu-1}(\mu = 1, \ldots, g)$. The dimension of this space is equal to $\deg D' + 1$, since any such integral is represented by

$$t(p) = c_o + \sum_{j=1}^{n} \sum_{r=1}^{\alpha_j} c_{jr} t^r_{q_j}(p), \tag{3.18}$$

where c_{jr} are arbitrary complex constants and $t^r_{q_k}$ are normalized Abelian integrals of the second kind (2.15). In order that the integral (3.18) be a function belonging to the space $L(D)$, it is necessary and sufficient that the periods along all cycles $k_{2\mu}(\mu = 1, \ldots, g)$ are zeros, namely that

$$\int_{k_{2\mu}} dt = \sum_{j=1}^{n} \left(\sum_{r=1}^{\alpha_j} c_{jr} \int_{k_{2\mu}} dt^r_{q_j}(p) \right) = 0, \qquad \mu = 1, \ldots, g, \tag{3.19}$$

and that $t(p)$ have zeros at the points of D'',

(3.20)
$$\left(c_o + \sum_{j=1}^{n}\left(\sum_{r=1}^{\alpha_j} c_{jr} t_{q_j}^{r}(p_k)\right)\right)^{(s)} = 0, \qquad s = 0,\ldots,\beta_k - 1,$$
$$k = 1,\ldots,m.$$

Therefore,

(3.21)
$$\dim L(D) = \deg D' - r + 1,$$

where r is a rank of the system (3.19) and (3.20).

Rewrite (3.19) with regard to (3.14),

(3.22)
$$\sum_{j=1}^{n}\left(\sum_{r=1}^{\alpha_j} \frac{c_{jr}}{(r-1)!}\frac{d^r w_\mu(q_j)}{dz_j^r}\right) = 0, \mu = 1,\ldots,g.$$

Consider the homogeneous system conjugate to the system (3.20) to (3.22) (see Figure 6),

$$\begin{array}{cccccccc}
0 & \frac{dw_1(q_1)}{dz} & \cdots & \frac{1}{(\alpha_1-1)!}\frac{d^{\alpha_1} w_1(q_1)}{dz^{\alpha_1}} & \cdots & \frac{1}{(\alpha_n-1)!}\frac{d^{\alpha_n} w_1(q_n)}{dz^{\alpha_n}} & \lambda_1 \\
\cdots & \cdots & \cdots & \cdots & \cdots & \cdots & \cdots \\
0 & \frac{dw_g(q_1)}{dz} & \cdots & \frac{1}{(\alpha_1-1)!}\frac{d^{\alpha_1} w_g(q_1)}{dz^{\alpha_n}} & \cdots & \frac{1}{(\alpha_n-1)!}\frac{d^{\alpha_n} w_g(q_n)}{dz^{\alpha_n}} & \lambda_g \\
1 & t_{q_1}^{1}(p_1) & \cdots & t_{q_1}^{\alpha_1}(p_1) & \cdots & t_{q_n}^{\alpha_n}(p_1) & \varepsilon_{11} \\
\cdots & \cdots & \cdots & \cdots & \cdots & \cdots & \cdots \\
0 & (t_q^1(p_1))^{(\beta_1-1)} & \cdots & (t_{q_1}^{\alpha_1}(p_1))^{(\beta_1-1)} & \cdots & (t_{q_n}^{\alpha_n}(p_1))^{(\beta_1-1)} & \varepsilon_{1,\beta_1-1} \\
1 & \cdots & \cdots & \cdots & \cdots & \cdots & \cdots \\
\cdots & \cdots & \cdots & \cdots & \cdots & \cdots & \cdots \\
0 & (t_{q_1}^1(p_m))^{(\beta_m-1)} & \cdots & (t_{q_1}^{\alpha_1}(p_m))^{(\beta_m-1)} & \cdots & (t_{q_n}^{\alpha_n}(p_m))^{(\beta_m-1)} & \varepsilon_{m,\beta_m^{-1}} \\
C_0 & C_{11} & \cdots & C_1\alpha_1 & \cdots & C_n\alpha_n & \cdots
\end{array}$$

Fig. 6. The matrix of the systems (3.20)–(3.22) and (3.23)

$$\sum_{\mu=1}^{g}\frac{\gamma_\mu}{(r-1)!}\frac{d^r w_\mu(q_j)}{dz_j^r} + \sum_{k=1}^{m}\sum_{s=0}^{\beta_k-1}\varepsilon_{ks}\left(t_{q_j}^{r}(p_k)\right)^{(s)} = 0,$$

(3.23)
$$r = 1,\ldots,\alpha_j, \qquad j = 1,\ldots,n, \qquad \varepsilon_{11} + \cdots + \varepsilon_{m1} = 0.$$

Substituting Eq. (3.15) in Eq. (3.23), we obtain

$$\sum_{\mu=1}^{g}\gamma_\mu\frac{d^r w_\mu(q_j)}{dz_j^r} - \sum_{k=1}^{m}\sum_{s=0}^{\beta_k-1}\varepsilon_{ks}\left(\frac{d^r \omega_{q_0 p_k}(q_j)}{dz_j^r}\right)^{(s)} = 0,$$

(3.24)
$$r = 1,\ldots,\alpha^j, \qquad j = 1,\ldots,n, \qquad \varepsilon_{11} + \cdots + \varepsilon_{m1} = 0.$$

Here q_0 is an arbitrary point of M where chosen branches of all integrals t_{q_k} are equal to zero. Each solution $(\gamma_\mu, \varepsilon_{kr})$ of Eq. (3.24) generates the differential

$$\omega = \sum_{\mu=1}^{g} \gamma_\mu dw_\mu(q) - \sum_{k=1}^{m} \sum_{s=0}^{\beta_k - 1} \varepsilon_{ks} (d\omega_{q_0 p_k}(q))^{(s)}. \tag{3.25}$$

Let us remember that the symbols $(f)^{(s)}$ correspond to derivatives of order s with respect to some fixed local coordinates of points p_k.

This differential is a multiple of the divisor D' and a multiple of the divisor $-D''$. Equation (3.24) means that the sum of residues of (3.25) is equal to zero. Therefore, $\omega \in H(D)$ and Eq. (3.24) have $\dim H(D)$ solutions. Hence the rank of this equation is

$$r = \deg D'' + g - \dim H(D). \tag{3.26}$$

Substituting Eq. (3.26) into Eq. (3.21), we obtain (3.22).

Other proofs of this famous theorem and some of its corollaries will be given below.

§4 Cauchy-type Integrals

The kernel of a Cauchy-type integral on the plane,

$$M(t,z)dt = \frac{dt}{t-z}$$

is an Abelian differential of the third kind with respect to a variable t having poles of the first order at the points $t = z, \infty$, with residues ± 1, respectively. One is an analytic function with respect to z having a pole of the third order at $z = t$ and zero at $z = \infty$. As is easily seen, these properties provide all principal applications of Cauchy-type integrals.

On a Riemann surface, it is also possible to construct some differential forms of two variables allowing the consideration of Cauchy-type integrals.[2] Here we use a construction proposed in the papers by Gusman & Rodin [a], and K.L. Volkovyski [a]. It is clear that the basis of any such construction has to be an Abelian integral of the third kind. It is important that the forms obtained depend analytically on parameters.

Apply the bilinear relations (3.3) and suppose that $df = d\omega_{q_0 q}, dg = d\omega_{p_0 p}$ (see (2.16)). The cut s is drawn from point q_0 to point q as in §3 (see Figure 5). Then

$$L(f,g) = 0, \qquad \sum_M \operatorname{res} \omega_{q_0 q} d\omega_{p_0 p} = -\omega_{q_0 q}(p) + \omega_{q_0 q}(p), \qquad \Delta f = \Delta \omega_{q_0 q} = 2\pi i.$$

Therefore, we obtain the relation

$$\omega_{p_0 p}(q) - \omega_{p_0 p}(q_0) = \omega_{q_0 q}(p) - \omega_{q_0 q}(p_0) \tag{4.1}$$

[2]Behnke & Stein [a].

involving an analytic dependence of both sides of this equation on both pairs of variables.

The situation is more complicated for the real normalization of periods of Abelian differentials of the third kind (2.16). Let $df = d\Omega_{q_0 q}$, $dg = d\Omega_{p_0 p}$. Then we have

$$\int_{s} \Delta f dg = 2\pi i \int_{q_0}^{q} d\Omega_{p_0 p} = 2\pi i[\Omega_{p_0 p}(q) - \Omega_{p_0 p}(q_0)],$$

$$\sum_{M} \operatorname{res} \Omega_{q_0 q} d\Omega_{p_0 p} = -\Omega_{q_0 q}(p_0) + \Omega_{q_0 q}(p).$$

Taking into account (3.12), we obtain

$$L(\Omega_{q_0 q}, \Omega_{p_0 p}) = \sum_{\mu=1}^{g} \Big(\int_{k_{2\mu-1}} d\Omega_{q_0 q} \int_{k_{2\mu}} d\Omega_{p_0 p}$$

$$- \int_{k_{2\mu}} d\Omega_{q_0 q} \int_{k_{2\mu-1}} d\Omega_{p_0 p} = (2\pi i)^2 \sum_{\mu=1}^{g} \operatorname{Im} \int_{q_0}^{q} d\theta_{2_{\mu-1}} \operatorname{Im} \int_{p_0}^{p} d\theta_{2\mu}$$

$$- \operatorname{Im} \int_{q_0}^{q} d\theta_{2\mu} \operatorname{Im} \int_{p_0}^{p} d\theta_{2\mu-1} \Big).$$

Finally we obtain

$$\begin{aligned} (4.2) \quad \Omega_{p_0 p}(q) - \Omega_{p_0 p}(q_0) &= \Omega_{q_0 q}(p) - \Omega_{q_0 q}(p_0) \\ &- 2\pi i \sum_{\mu-1}^{g} \left(\operatorname{Im} \int_{q_0}^{q} d\theta_{2\mu-1} \operatorname{Im} \int_{p_0}^{p} d\theta_{2\mu} - \operatorname{Im} \int_{q_0}^{q} d\theta_{2\mu} \operatorname{Im} \int_{p_0}^{p} d\theta_{2\mu-1} \right). \end{aligned}$$

Let

$$(4.3) \qquad M^*(p, q)dp = \partial_p[\omega_{p_0 p}(q) - \omega_{p_0 p}(q_0)],$$

$$(4.4) \qquad M_*(p, q)dp = \partial_p[\Omega_{p_0 p}(q) - \Omega_{p_0 p}(q_0)].$$

Here

$$(4.5) \quad \partial_p = \frac{\partial}{\partial z(p)} \wedge dz(p) = \frac{1}{2}\Big(\frac{\partial}{\partial x} - i\frac{\partial}{\partial y}\Big) \wedge (dx + idy), \qquad z = z(p) = x + iy.$$

Points p_0 and q_0 are fixed in the following.

We shall make sure now that both these forms can be used as kernels of Cauchy-type integrals of the surface M.

It is obvious from the definitions that both forms are analytic with respect to q. One can verify that they are analytic with respect to the first variable, too. For the form (4.3), it follows from Eq. (4.1).

For the form (4.4), we have from Eq. (4.2)

$$(4.6)\qquad M_*(p,q)dp = d\Omega_{q_0q}(p) - \pi\sum_{\mu=1}^{g}\left\{\left(\operatorname{Im}\int_{q_0}^{q} d\theta_{2\mu-1}\right)d\theta_{2\mu} - \left(\operatorname{Im}\int_{q_0}^{q} d\theta_{2\mu}\right)d\theta_{2\mu-1}(p)\right\},$$

which involves the analyticity with respect to p. To produce (4.6), we used the relation

$$\partial(\operatorname{Im} f(q)) = \partial\left(\frac{f(z)-\overline{f(z)}}{2i}\right) = \frac{1}{2i}df,$$

which is true for any analytic function.

Both forms (4.3) and (4.4) have the residues ± 1 at the points $p = q, q_0$, respectively. Equations (4.1) and (4.6) show that these values are independent of p_0.

As functions of q, both forms (4.3) and (4.4) are Abelian integrals of the second kind with poles of the first order at $q = p$. The periods of these integrals are equal to (cf. Eq. (3.12) and (3.13))

$$(4.7)\qquad \begin{aligned} l_\mu(p)dp &= \int_{k_{2\mu}} d_q M^*(p,q)dp = \partial_p 2\pi i\int_{p_0}^{p} dw_\mu = 2\pi i\, dw_\mu(p),\\ &\int_{k_{2\mu-1}} d_q M^*(p,q)dp = 0, \qquad \mu = 1,\dots,g. \end{aligned}$$

(4.8)

$$L_j(p)dp = \int_{k_j} d_q M_*(p,q) = \partial_p 2\pi i \operatorname{Im}\int_{p_0}^{p} d\theta_j = \pi d\theta_j(p), \qquad j = 1,\dots,2g.$$

Cut the surface M along the cycles of the canonical homology basis. We obtain the polygon $\widehat{M}$. The kernels (4.3) and (4.4) are single-valued in $\widehat{M}$. We choose their single-valued branches on M by the condition

$$(4.9)\qquad M^*(p,q_0) = M_*(p,q_0) = 0.$$

If L is some sufficiently smooth curve on $\widehat{M}$ and $\varphi(p)$ is an H-continuous function on L, we can consider the Cauchy-type integrals (see Appendix 4)

$$(4.10)\qquad f(q) = \frac{1}{2\pi i}\int_L \varphi(p)M^*(p,q)dp,$$

$$(4.11)\qquad F(q) = \frac{1}{2\pi i}\int_L \varphi(p)M_*(p,q)dp.$$

The Plemelj–Sokhotsky formulae (see [9], [18]),

$$
(4.12)\qquad
\begin{aligned}
f^{\pm}(q) &= \pm\frac{1}{2}\varphi(q) + \frac{1}{2\pi i}\int\limits_L \varphi(p)M^*(p,q)dp,\\
F^{\pm}(q) &= \pm\frac{1}{2}\varphi(q) + \frac{1}{2\pi i}\int\limits_L \varphi(p)M_*(p,q)dp, \qquad q\in L,
\end{aligned}
$$

are valid on L. Here the signs $\pm$ correspond to the limit values to the right and left of the contour, and the integrals (4.12) are taken in the sense of the Cauchy principal values. In general, both functions $f(q)$ and $F(q)$ are multivalued on the surface M, and their periods are equal to

$$
(4.13)\qquad f_\mu = \int\limits_{k_{2\mu}} df = \int\limits_L \varphi(p)dw_\mu(p), \qquad \int\limits_{k_{2\mu-1}} df = 0, \qquad \mu = 1,\dots,g.
$$

$$
(4.14)\qquad F_j = \int\limits_{k_j} dF = \frac{1}{2i}\int\limits_L \varphi(p)d\theta_j(p), \qquad j = 1,\dots,2g.
$$

We shall consider the single-valued branches of these integrals obtained by the cutting of M. These branches are fixed by the conditions

$$
(4.15)\qquad f(q_0) = F(q_0) = 0.
$$

The considered functions (more exactly, the branches) have finite discontinuities along the sections, and the values of jumps are equal to the periods of integrals along dual cycles (see Figure 7). In some cases it is convenient to use Cauchy-type integrals single-valued on M. We proceed to the construction of the single-valued kernel of a Cauchy-type integral. This single-valuedness is achieved by the introduction of added poles.

Let $\delta = \sum \alpha_k q_k = 0$ be a divisor such that $\dim H(\delta) = 0$ (see §3C). Evaluate $\deg\delta$. By the Riemann–Roch theorem, $\dim H(\delta) = \dim L(\delta) + g - 1 - \deg\delta$. If $\deg\delta < g$, $\dim H(\delta) > 0$. Hence $\deg\delta \geqslant g$. Consider a complex-normalized Abelian integral of the second kind which is a multiple of the divisor $-\delta$,

$$
(4.16)\qquad t_\delta(q) = \sum_k \sum_{j=1}^{\alpha_k} c_{kj} t^j_{q_k}(q)
$$

(we omit a symbol for a local parameter in $t^j_{q_{kz}}$ in (2.14)). We do not fix any coefficients of (4.16). The periods of this integral are

$$
l_{\delta\mu} = \int\limits_{k_{2\mu}} dt_\delta = \sum_{k'}\sum_{j=1}^{\alpha_{k'}} c_{k'j}\int\limits_{k_{2\mu}} dt^j_{q_{k'}}, \qquad \mu = 1,\dots,g.
$$

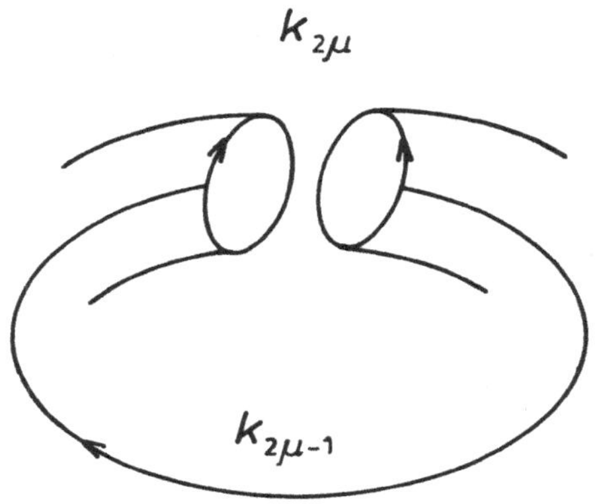

Figure 7

The number of solutions of the homogeneous system of equations

$$\sum_{k'}\sum_{j=1}^{\alpha_{k'}} c_{k'j} \int_{k_{2\mu}} dt^j_{q_{k'}} = 0, \qquad \mu = 1, \dots, g.$$

is equal to $\dim L(\delta) - 1$ (at the cost of a constant). By the Riemann–Roch theorem, we have

$$\dim L(\delta) = \deg \delta - g + 1,$$

since $\dim H(\delta) = 0$. Hence it appears that the rank r of the matrix

$$\left(\int_{k_{2\mu}} dt^j_{qk'}\right) \qquad \mu = 1, \dots, g; \quad j = 1, \dots, \alpha_{k'}, \quad k' = 1, \dots$$

is equal to $r = \deg \delta - (\deg \delta - g + 1 - 1) = g$. Thus, the nonhomogeneous system

$$\sum_{k'}\sum_{j} c_{k'j} \int_{k_{2\mu}} dt^j_{q_{k'}} = l_\mu, \qquad \mu = 1, \dots, g \tag{4.17}$$

is solvable for any right-hand side. Let $l_\mu = l_\mu(p)$ be periods of the kernel of (4.3) determined by the relations (4.7). We obtain the values of the coefficients $c_{kj} = c_{kj}(p)$ from Eq. (4.17). Consider the form

$$M(p,q)\,dp = M^*(p,q)\,dp - \sum_{k}\sum_{j=1}^{\alpha_k} c_{kj}(p) t^j_{q_k}(q)\,dp. \tag{4.18}$$

The form $M(p,q)dp$ has conserved its poles $p = q$ and $p = q_0$ and has added new poles determined by the divisor δ. This divisor is called the characteristic divisor of the kernel. It is obvious that the kernel (4.18) is single-valued with respect to both variables.

When $\deg \delta = g$, we designate the divisor δ by a minimal divisor. For such a divisor,

$$\dim L(\delta) = g - g + 1 = 1,$$

i.e., there exists no analytic function different from the constant which is a multiple of $-\delta$.

Consider in detail the case in which the divisor has no multiple points, $\delta = \sum_{k=1}^{g} q_k$. The equation (4.17) takes the form (from Eq. (3.14))

$$-\sum_{k=1}^{g} c_k \frac{dw_\mu(q_k)}{dz_k(q)} = \frac{dw_\mu(p)}{dz(p)}, \qquad \mu = 1, \ldots, g,$$

where $z_k(q)$ are fixed local parameters in the neighbourhoods of the points q_k. It is obvious that in this case the solutions of the system are Abelian differentials of the first kind of the variable p,

$$c_k(p)dp = dZ_\mu(p),$$

determined by the conditions

$$\frac{dZ\mu(q_\nu)}{dz_\nu(q)} = -\delta_{\mu\nu}, \qquad \mu, \nu = 1, \ldots, g, \tag{4.19}$$

where dz_j are local coordinates described above. It is clear that the differentials dZ_k are linearly independent and form the basis of the space of Abelian differentials of the first kind, determined by the divisor δ. In this case we have

$$M(p,q)dp = M^*(p,q)dp - \sum_{k=1}^{g} t_{q_k}(q) dZ_k(p). \tag{4.20}$$

The behaviour of this kernel is well illustrated by Figure 8. Solid lines correspond to poles, and broken lines correspond to zeros of the kernel. The diagonal lines correspond to the pole $p = q$, the horizontal solid line to the pole $p = q_0$. The vertical solid lines correspond to poles $q = q_k$, and the horizontal broken lines correspond to zeros, $M(q_k, q) = 0$. These relations follow from Eq. (4.19) and from the relations

$$M^*(q_k, q) = -t_{q_k}(q), \qquad k = 1, \ldots, g. \tag{4.21}$$

Finally, the vertical broken line $M(p, q_0) = 0$ is provided by a corresponding choice of branches of all integrals in Eq. (4.20),

$$M^*(p, q_0) = -t_p(q_0) = 0. \tag{4.22}$$

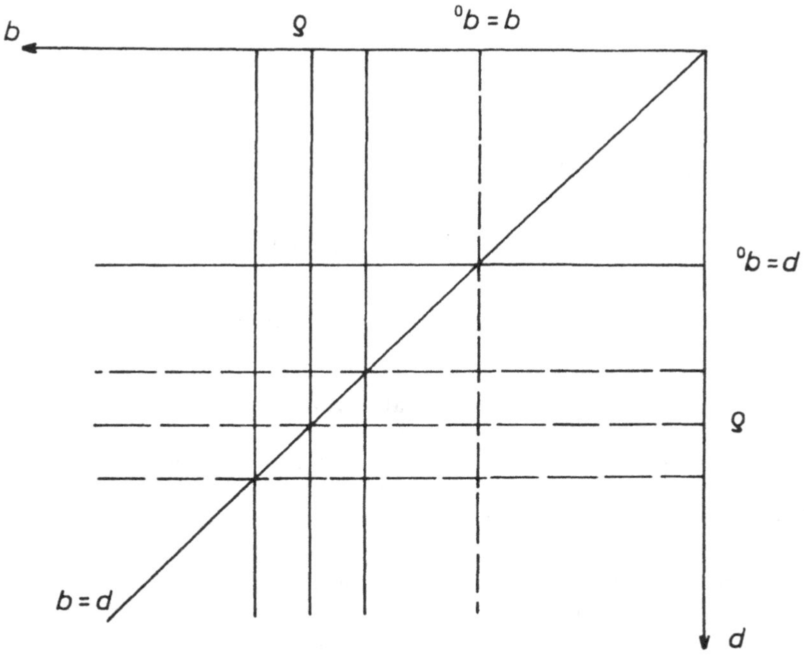

Figure 8

Note that at solid nodes the kernel is regular and equal to zero. It can be shown in the following manner. At the points $(q_k, q_k), k = 1, \ldots, g$, the kernel is zero (from Eq. (4.20) and (4.21)). $M(q_k, q_0) = 0, (k = 0, \ldots, g)$, because of (4.22).

The Cauchy-type integral

$$F(q) = \frac{1}{2\pi i} \int_L \varphi(p) M(p, q) dz(p) \tag{4.23}$$

is analytic on M and possesses the jump $\varphi(p)$ on L and poles at the points q_k. The principal parts of these poles in local coordinates fixed in (2.14) are equal to

$$-\frac{1}{2\pi i} \int_L \varphi(p) dZ_k(p), \qquad k = 1, \ldots, g. \tag{4.24}$$

If $\psi(p) dz(p)$ is a differential form on L, the Cauchy-type integral

$$f(q) = \frac{1}{2\pi i} \int_L \psi(p) dz(p) \wedge M(q, p) dz(q) \tag{4.25}$$

is a piecewise-analytic differential on M, being a multiple of the divisor δ and with a pole of the first order at the point q_0 (called a polar point of the kernel) with a residue

$$\operatorname{res}_{q_0} f = -\frac{1}{2\pi i}\int_L \psi(p)dz(p). \tag{4.26}$$

The Plemelj–Sokhotsky formulae have the form

$$\begin{aligned} F^{\pm}(q) &= \pm\frac{1}{2}\varphi(q) + \frac{1}{2\pi i}\int_L \varphi(p)M(p,q)dz(p), \\ f^{\pm}(q) &= \mp\frac{1}{2}\psi(q)dz(q) + \frac{1}{2\pi i}\int_L \psi(p)dz(p)\wedge M(q,p)dz(q), \end{aligned} \tag{4.27}$$

since the principal part of the diagonal singularity of the kernel changes the sign when the variables p and q are transposed.

Of course, above we have to assume that L is sufficiently smooth (for example, that L is a Liapounov contour) and that φ and ψ are H-continuous on L. This condition can be relaxed as it is made in the plane case.

Note. The condition $H(\delta) = 0$ is not very strong. Namely, choose $g-1$ arbitrary points of the surface M. There is a single Abelian differential of the first kind with zeros at these points. Each of the $g-1$ zeros remaining of this differential and only those can be chosen as a g-th point of the divisor δ.

§5 The Riemann Problem. Number of Solutions

Let M be a closed Riemann surface of genus g and let L be a contour without self-intersections on M, consisting of a finite number of sufficiently smooth curves dividing the surface into two domains $T^{\pm}$ and T^{-}. On the contour L we define a Hölder-continuous function $G(p)$ different from zero at every point of L.

The Riemann boundary problem. Determine in the domains $T^{\pm}$ functions $F^{\pm}(q)$ holomorphic and continuous up to the boundary satisfying the boundary condition on L,

$$F^{+}(p) = G(p)F^{-}(p), \qquad p \in L. \tag{5.1}$$

The problem (5.1) was formulated by L.I. Volkovyskii [a] at the Soviet conference on the theory of functions of complex variables in 1957. The first results in this area were reported by the author at these conferences in 1957–58 [c,d] and published in 1957–60 [a]–[e]. Simultaneously and independently, the same results were obtained by Koppelman [a,b]. An equivalence was established between the problem (5.1) and the main theorems of the theory of algebraic functions—Abel's and Riemann–Roch theorems. Later, in 1962, Chibricova [b] established the relation between this problem and the Jacobi inversion problem. A very convenient tool for an investigation of the problem (5.1), it presents additive and multiplicative multivalued solutions. Such functions were introduced by the author in 1961–62 [f,j,k]. Using multivalued Cauchy-type integrals, Abdulaev [a,b,c] in 1963 established the equivalence between the Jacobi inverse problem, Riemann theta-functions, and the solvability of the Riemann problem (5.1), and obtained solvability conditions for the index $0 < \kappa < g$. These investigations were added in 1970–71 by Zverovich [a–d] (see also his expository paper [26]), who noted that for every Riemann problem an equivalent one of the standard form (§8) can be constructed. It provides the possibility of constructively solving the problem. Grothendieck [a] established connections between the Riemann problem on the plane and complex linear bundles. In 1963 Röhrl [c] established these connections for the case of compact Riemann surfaces. He used modern means of algebraic topology to study this area (see Chapter 2).

Note 1. The contour L can divide or cannot divide a surface on two domains (for example, a cyclic section on a torus). In the latter case, L can be added to a dividing contour by a necessary number of curves. On the added curves we assume $G \equiv 1$.

Note 2. It is possible to demand that $F^{\pm}(q)$ be analytic functions which are multiples of the divisors $h_{\pm}$ in the domains $T^{\pm}$,

$$(F^{\pm}) \geqslant h_{\pm} \quad \text{in} \quad T^{\pm}. \tag{5.2}$$

Let $h_{\pm}(q)$ be analytic functions in $T^{\pm}$ whose divisors are $h_{\pm}$, $(h_{\pm}(q)) = h_{\pm}$. The existence of such functions is proved in §14A. Then a solution can be represented by

$$F^{\pm}(q) = h_{\pm}(q) F_0^{\pm}(q).$$

It is clear that $F^{\pm}(q)$ is a solution of the Riemann problem

$$F_0^{+}(p) = G_0(p) F_0^{-}(p), \qquad G_0(p) = h_{+}^{-1}(p) G(p) h_{-}(p). \tag{5.3}$$

Such statement of the problem is sometimes convenient.

Note 3. It is sufficient to demand that L be a Liapounov contour, i.e. that it possesses a Hölder-continuous rotating tangent. It provides applicability of the Plemelj–Sokhotsky formulae sufficient for our constructions. Constraints on L and $G(p)$ can be reduced (a rectifiable curve, a summable function). But effects generated by this generalization are not specific for Riemann surfaces and coincide with the plane case $g = 0$ (see, for example, [5]).

Note 4. Here we do not consider the case where $G(p)$ has zeros or jump discontinuities (or the equivalent case of an open or self-intersecting contour). These problems were studied by the author in [e]. This case makes no principal contribution in comparison with the plane case.

Along with problem (5.1) we consider a conjugate problem

$$f^{+}(p) = \frac{1}{G(p)} f^{-}(p) \tag{5.4}$$

for holomorphic differentials $f^{\pm}(p)\,dz(p)$ in $T^{\pm}$.

First consider the simplest problem,

$$F^+(p) - F^-(p) = g(p). \tag{5.5}$$

THEOREM 5.1. *For problem (5.5) to be solvable, it is necessary and sufficient that*

$$\int_L g(p)dZ_\mu(p) = 0, \qquad \mu = 1, \ldots, g \tag{5.6}$$

be valid, where $dZ_\mu(\mu = 1, \ldots, g)$ is any complex basis of the space of Abelian differentials of the first kind.

The necessity of Eq. (5.6) is obtained directly. It is sufficient to multiply (5.5) by an Abelian differential of the first kind and to integrate over L. Both integrals on the left-hand side are equal to zero by the Cauchy integral theorem.[3]

In order to show sufficiency of (5.6), write a solution of the problem (5.4) having, in general, poles at the points of the characteristic divisor δ

$$F(q) = \frac{1}{2\pi i} \int_L g(p)M(p,q)dz(p). \tag{5.7}$$

Here $M(p,q)$ is the Cauchy kernel determined by Eq. (4.20). Then, from (4.24), it follows that (5.7) is a holomorphic solution if Eq. (5.6) is valid.

As always [9,18], the index of the boundary condition (the Cauchy index) is called the integer

$$\kappa = \operatorname{ind}_L G = \frac{1}{2\pi}\Delta_L \arg G(p) = \frac{1}{2\pi i}\int_L d\ln G(p). \tag{5.8}$$

Note 5. For problem (5.2), the index of $G(p)$, as is easy to see, is equal to

$$\tilde{\kappa} = \kappa + \deg(h_+ + h_-). \tag{5.9}$$

Therefore, permitting some number of poles of a solution, we increase the Cauchy index of the problem. This may be very convenient.

THEOREM 5.2. *The problem (5.1) is solvable when $\kappa \geqslant g$.*

We assume that a solution of (5.1) has the form

$$F(q) = \frac{1}{2\pi i}\int_L \varphi(p)M(p,q)dz(p) + C, \tag{5.10}$$

[3]It is sufficient to triangulate the domain and to integrate over boundaries of triangles.

where C is any constant. Using (4.27) and (5.1), we obtain the singular integral equation

$$\frac{1+G(q)}{2}\varphi(q)+\frac{1-G(q)}{2\pi i}\int_L \varphi(p)M(p,q)dz(p)=[G(q)-1]C \tag{5.11}$$

of the index κ [9,18].[4] A number of solutions of the homogeneous equation (5.11) is equal to $s=\kappa+s'$, where s' is a number of solutions of the conjugate equation. If Eq. (5.11) is solvable for $C\neq 0$, the number of its solutions is $\tilde{s}=s+1$. In the opposite case, $\tilde{s}=s$. We have for $s'>0:\tilde{s}\geqslant\kappa+s'>g$; for $s'=0:\tilde{s}=\kappa+1>g$.

Every solution of Eq. (5.11) determines some solution of the problem of (5.1) by formula (5.10). In general, this solution possesses poles at the points of the characteristic divisor whose principal parts are determined by Eq. (4.24). A rank of the principal parts matrix

$$\left\|\int_L \varphi_j(p)dZ_k\right\|, \qquad j=1,\ldots,\tilde{s}; \qquad k=1,\ldots g,$$

where φ_j are solutions of (5.11) less than or equal to g; consequently, there exists a zero combination of matrix rows corresponding to a regular solution of the problem (5.1).

From Theorem 5.2 it follows that every problem (5.1) has a solution if allowed to admit $g-\kappa(\kappa\leqslant g)$ poles (see Note 2).

Note that calculating an increment in the argument of both sides of Eq. (5.1), we obtain the relation

$$\frac{1}{2\pi i}\int_L d\ln F^+ - \frac{1}{2\pi i}\int_L d\ln F^- = \kappa.$$

It means that a divisor (F) of any solution $F^\pm$ has a degree $\deg(F)=\kappa$. From this we see that for $\kappa<0$ the problem (5.1), as in the plane case, has no holomorphic solutions.

THEOREM 5.3. *The difference of a number 1 of the holomorphic solutions of the problem (5.1) and a number h of the holomorphic solutions of the conjugate problem (5.4) is equal to*

$$l-h=\kappa-g+1 \tag{5.12}$$

(dimensions are complex).

Let $F_0(q)$ be a solution which, in general, has poles. Then by substituting the value $F_0^+(p)/F_0^-(p)$ in (5.1) instead of $G(p)$, we obtain that

$$\frac{F^+(p)}{F_0^+(p)}=\frac{F^-(p)}{F_0^-(p)} \qquad \text{on} \quad L,$$

[4] See Appendix 2.

and hence

$$H(q) = \begin{cases} F^+(q)/F_0^+(q) & q \in T^+ \\ F^-(q)/F_0^-(q) & q \in T^- \end{cases}$$

is the analytic function on M which is the multiple of the divisor $-(F_0)$. It is clear that the space of holomorphic solutions of the problem (5.1) is isomorphic to the space $L(F_0)$ of functions being multiples of the divisor $-(F_0)$ (see §3C). On the other hand, from Eq. (5.4) we have that

$$f^+(p)F_0^+(p)dz(p) = f^-(p)F_0^-(p)dz(p) = \omega(p)$$

is a differential form analytic on M, which is a multiple of the divisor (F_0). Therefore, the space of holomorphic solutions of the problem (5.4) is isomorphic to the space $H(F_0)$ of differentials which are multiples of (F_0). Hence, $1 = \dim L(F_0), h = \dim H(F_0)$, and our statement follows from the Riemann–Roch theorem (§3C). From Theorem 5.3 and *Note 2* we obtain the following theorem.

THEOREM 5.3′. *The difference of a number l' of solutions the problem (5.1) which are multiples of the divisor $-D$ and a number h' of solutions of the problem (5.4) which are multiples of the divisor D is equal to*

$$l' - h' = \kappa + \deg D - g + 1. \tag{5.12'}$$

Note 6. Below we shall get another proof of Theorem 5.3 without the Riemann–Roch theorem. The Riemann–Roch theorem follows from Theorem 5.3′ when $G = 1$.

Now we finally expect singular integral equations which correspond to problems (5.1) and (5.4), and we obtain the other proof of Theorem 5.3.

Let δ be a minimal characteristic divisor of some kernel (4.20) and q_0 be its polar point. Let $\Delta = \delta - q_0$. For simplicity, we assume that the points of Δ do not belong to L. Let $\Delta = \Delta^+ + \Delta^-$, where $\Delta^\pm$ are divisors whose points belong to the domains $T^\pm$, respectively. Denote by $\Delta^\pm(q)$ analytic functions in $T^\pm$ whose zeros and poles are described by the divisors $\Delta^\pm$ (i.e., these functions have zeros at the points of δ and a pole of the first order at $q = p_0$). We shall prove the existence of such functions in §14A.

We shall look for solutions of the problems (5.1) and (5.4) of the form

$$\begin{aligned} F^\pm(q) &= \tilde{F}^\pm(q)\Delta^\pm(q), \\ f^\pm(q) &= \frac{\tilde{f}^\pm(q)}{\Delta^\pm(q)}. \end{aligned} \tag{5.13}$$

Then (5.1) is reduced to the problems

$$\tilde{F}^+(p) = \tilde{G}(p)\tilde{F}^-(p), \qquad (\tilde{F}) + \Delta \geqslant 0 \tag{5.14}$$

$$\tilde{f}^+(p) = \frac{1}{\tilde{G}(p)}\tilde{f}^-(p), \qquad (\tilde{f}) \geqslant \Delta \tag{5.15}$$

$$\tilde{G}(p) = G(p)\frac{\Delta^-(p)}{\Delta^+(p)}. \tag{5.16}$$

The Cauchy index of the function $\widetilde{G}(p)$ is equal to

$$\tilde{\kappa} = \kappa - g + 1. \tag{5.17}$$

LEMMA. *All solutions of problems (5.14) and (5.15) are represented in the form*

$$\widetilde{F}^{\pm}(q) = \frac{1}{2\pi i} \int_L \varphi(p) M(p, q) dz(p), \tag{5.18}$$

$$\tilde{f}^{\pm}(q) = -\frac{1}{2\pi i} \int_L \psi(p) M(q, p) dz(p). \tag{5.19}$$

In fact, let $H^{\pm}(q)$ and $h^{\pm}(q)$ be analytic functions and differentials defined in $T^{\pm}$ which are multiples of the divisors $-\Delta^{\pm}$ and $\Delta^{\pm}$, respectively. Assume that

$$\begin{aligned} \varphi(p) &= H^{+}(p) - H^{-}(p) \\ \psi(p) &= h^{+}(p) - h^{-}(p) \end{aligned} \quad \text{on} \quad L \tag{5.20}$$

and consider the integrals (5.18) and (5.19) with densities (5.20). It is obvious that $\widetilde{F}^{\pm}(q) - H^{\pm}(q)$ is an analytic function on M which is a multiple of the divisor $-\delta$. With regard to the condition $\dim H(\delta) = 0$ (see §4), we obtain from the Riemann–Roch theorem that

$$\dim L(\delta) = g - g + 1 = 1.$$

Because $F^{\pm}(q_0) = 0$, we have $F^{\pm}(q) - H^{\pm}(q) \equiv \text{const} = 0$. The value $\tilde{f}^{\pm}(q) - h^{\pm}(q)$ is an Abelian differential having a single pole of the first order and being a multiple of δ. Because $\dim H(\delta) = 0$, such a differential is identically equal to zero. Using the representations (5.18), (5.19), and Eq. (4.27), we reduce the problems (5.14) and (5.15) to the singular integral equations

$$\frac{1 + \widetilde{G}(q)}{2} \varphi(q) + \frac{1 - \widetilde{G}(q)}{2\pi i} \int_L \varphi(p) M(p, q) dz(p) = 0 \tag{5.21}$$

$$q \in L$$

$$\frac{1 + \widetilde{G}^{-1}(q)}{2} \psi(q) - \frac{1 - \widetilde{G}^{-1}(q)}{2\pi i} \int_L \psi(p) M(q, p) dz(p) = 0 \tag{5.22}$$

of index $\tilde{\kappa}$ conjugate to each other. The difference of the numbers of solutions of the homogeneous equations (5.21) and (5.22) is equal to $s - s' = \tilde{\kappa}$ [18]. Every solution of Eq. (5.21) determines a solution of the problem (5.14), and conversely. The same is valid for Eq. (5.22) and the problem (5.15). Hence $l = s, h = s'$. We have

$$l - h = \kappa - g + 1,$$

which concludes our proof. Therefore, formula (5.12) describing the index of the Riemann problem is equivalent to the Riemann–Roch theorem.

Note that in the case $\deg D > 2g-2$, $\dim H(D) = 0$, since there exists no differential having more than $2g-2$ zeros. On the other hand, for $\deg D < 0, \dim L(D) = 0$. Summing it with Theorem 5.2, we can conclude the following. Problem (5.1) has no holomorphic solution when $\kappa < 0$; for $0 \leqslant \kappa < g$ the number of solutions can vary from zero to $\kappa + 1$, for $g \leqslant \kappa \leqslant 2g-2$ one varies from $\kappa - g + 1$ to $\kappa + 1$; for $\kappa > 2g-2$ the problem (5.1) has $\kappa - g + 1$ solutions. These estimates can be made somewhat more accurate (see [26]). Note that for $\kappa = 0$ the problem (5.1) has a single solution (if the problem is solvable). This solution is determined up to a constant multiplier. The condition $F(s) = 1$, where s is an arbitrary point of M, normalizes a solution. But in some applications such normalization is impossible (Rodin [s]). The calculation of a normalized multiplier (in general, depending on certain parameters) can be difficult. For this purpose in particular, a symmetry group of the problem (see Chapter 6) can be used.

In conclusion, we examine the nonhomogeneous problem

$$F^+(p) = G(p)F^-(p) + g(p) \qquad \text{on} \quad L, \tag{5.23}$$

where $g(p)$ is a Hölder-continuous function. At first, we consider the problem

$$\Phi^+(p) - \Phi^-(p) = g(p), \qquad (\Phi) + D \geqslant 0, \tag{5.24}$$

where D is a preset divisor, $D \cap L = \emptyset$.

THEOREM 5.4. *In order that the problem (5.24) be solvable it is necessary and sufficient that the conditions*

$$\int_L g(p)\omega_j(p) = 0 \qquad j = 1, \ldots, \dim H(D) \tag{5.24'}$$

be valid. Here $\omega_j (j = 1, \ldots, \dim H(D))$ *is a basis of the space* $H(D)$.

Necessity of (5.24) is proved directly by multiplying (5.24) by $\omega_j(p)$ and by integration.

Consider the Cauchy-type integral

$$\tilde{\Phi}(q) = \frac{1}{2\pi i} \int_L g(p) M(p, q)\, dz(p) \tag{5.25}$$

with kernel (4.20) whose poles are denoted (in this case) by $s_1, \ldots, s_g, \delta = \sum_{\nu=1}^{g} s_\nu$. Let $D = D' - D''$, $D' = \sum_{j=1}^{n} \alpha_j q_j \geqslant 0, D'' = \sum_{k=1}^{m} \beta_k p_k \geqslant 0$. Consider the function

$$\Phi(q) = \tilde{\Phi}(q) + c_0 + \sum_{j=1}^{n} \sum_{r=1}^{\alpha_j} c_{jr} t_{q_j}^r(q) + \sum_{\nu=1}^{g} c_\nu t_{s_\nu}(q). \tag{5.26}$$

Here $t^r_{q_j}$ are Abelian integrals of the second kind (2.14), and (2.15) and

$$c_\nu = \frac{1}{2\pi i}\int_L g(p)dZ_\nu(p), \qquad \nu = 1,\dots,g. \tag{5.26'}$$

Because of (4.24), $\Phi(q)$ is regular at the points s_ν. $\Phi(q)$ is a solution of (5.24) if it is single-valued and a multiple of the divisor D''. Calculating the periods of $\Phi(q)$ along $k_{2\mu}$, we obtain by using (3.14) and (4.13) the single-valued conditions

$$\sum_{\nu=1}^{g} c_\nu \frac{dw_\mu(s_\nu)}{dz_\nu} + \sum_{j=1}^{n}\sum_{r=1}^{\alpha_j} \frac{c_{jr}}{(r-1)!}\frac{d^r w_\mu(q_j)}{dz_j^r} = 0, \qquad \mu = 1,\dots,g. \tag{5.27}$$

At the points of D'', we have

$$\begin{gathered}\widetilde{\Phi}^s(p_k) + \left(\sum_{\nu=1}^{g} c_\nu t_{s_\nu}(p_k)\right)^{(s)} \\ + \left(c_0 + \sum_{j=1}^{n}\sum_{r=1}^{\alpha_j} c_{jr} t^r_{q_j}(p_k)\right)^{(s)} = 0, \\ s = 0,\dots,\beta_k - 1; \qquad k = 1,\dots,m.\end{gathered} \tag{5.28}$$

Here (s) means differentiation with respect to some fixed local coordinate of p_k. Consider the homogeneous linear system conjugate to (5.27) and (5.28). We have (see (3.24))

$$\begin{gathered}\sum_{\mu=1}^{g} \gamma_\mu \frac{d^r w_\mu(q_j)}{dz_j^r} - \sum_{k=1}^{m}\sum_{s=0}^{\beta_k-1} \varepsilon_{ks}\left(\frac{d^r \omega_{q_0 p_k}(q_j)}{dz_j^r}\right)^{(s)} = 0, \\ r = 1,\dots,\alpha_j; \quad j = 1,\dots,n, \quad \varepsilon_{11} + \cdots + \varepsilon_{m1} = 0.\end{gathered} \tag{5.29}$$

Every solution $(\gamma_\mu, \varepsilon_{kr})$ of Eq. (5.29) generates the differential form

$$\omega = \sum_{\mu=1}^{g} \gamma_\mu dw_\mu - \sum_{k=1}^{m}\sum_{s=0}^{\beta_k-1} \varepsilon_{ks}(d\omega_{q_0 p_k}(q))^{(s)}, \qquad \omega \in H(D) \tag{5.30}$$

(see (3.25)). The solvability conditions for Eqs. (5.27) and (5.28) are

$$\sum_{\mu=1}^{g} \gamma_\mu \sum_{v=1}^{g} c_\nu \frac{dw_\mu(s_\nu)}{dz_\nu} + \sum_{k=1}^{m}\sum_{s=0}^{\beta_k-1} \varepsilon_{ks}\left[\Phi_{(p_k)} + \sum_{v=1}^{g} c_\nu t_{s_\nu}(p_k)\right]^{(s)} = 0 \tag{5.31}$$

for any solution $(\gamma_\mu, \varepsilon_{ks})$ of Eq. (5.29). We have

$$\begin{aligned}\Phi(p_k) + \sum_{\nu=1}^{g} c_\nu t_{s_\nu}(p_k) &= \frac{1}{2\pi i}\int_L \Big\{ g(p) M^*(p_1 p_k) dp \\ &- \sum_{\nu=1}^{g} t_{s_\nu}(p_k) Z'_\nu(p) dp + \sum_{\nu=1}^{g} t_{s_\nu}(p_k) Z'_\nu(p) dp \Big\} = \frac{1}{2\pi i}\int_L g(p) d\omega_{q_0 p_k}(p).\end{aligned}$$

The relations $\dim H(\delta) = 0$ and (4.19) involve the relation

$$\sum_{\nu} \frac{dw_\mu(s_\nu)}{dz_\nu} dZ_\nu(p) = -dw_\mu(p).$$

Thus we have

$$\sum_{\nu} c_\nu \frac{dw_\mu(s_\nu)}{dz_\nu} = -\frac{1}{2\pi i} \int_L g(p) dw_\mu(p), \qquad \mu = 1, \ldots g.$$

Eq. (5.31) takes the form

$$\int_L g(p)\omega(p) = 0, \qquad \omega \in H(D),$$

which is equivalent to (5.24).

THEOREM 5.5. *In order that the problem*

$$F^+(p) = G(p)F^-(p) + g(p), \qquad (F) + D \geqslant 0 \tag{5.23'}$$

be solvable, it is necessary and sufficient that the conditions

$$\int_L g(p) f_j^+(p) dz(p) = 0, \qquad j = 1, \ldots, h \tag{5.32}$$

be valid. Here $f_j^+(p)(j = 1, \ldots, h)$ *is a full system of solutions of the problem*

$$f^+(p)dz(p) = G^{-1}(p) f^-(p) dz(p), \qquad (fdz) \geqslant D. \tag{5.4'}$$

The necessity of (5.23) is proved directly as in Theorem 5.1.

Now let $F_0^\pm(q)$ be a solution of (5.1) having some poles. Substituting $G(p) = F_0^+(p)/F_0^-(p)$ into (5.23′), we reduce our problem to

$$\frac{F^+(p)}{F_0^+(p)} - \frac{F^-(p)}{F_0^-(p)} = \frac{g(p)}{F_0^+(p)}, \qquad (\frac{F}{F_0}) + D + (F_0) \geqslant 0. \tag{5.33}$$

The solvability conditions of (5.33) are

$$\int_L \frac{g(p)}{F_0^+(p)} \omega(p) = 0, \qquad \omega(p) \in H(D + (F_0)),$$

which coincides with (5.32).

§6 Inversion of Abelian Integrals and Abel's Theorem. Solvability of the Riemann Problem.

A. *Abel's Theorem*

On the surface M consider a divisor $D = \sum_k \alpha_k p_k, \deg D = 0$. When can we construct an analytic function $f(p)$ whose zeros and poles are determined by the divisor D, $(f) = D$?

Construct the complex normalized Abelian integral of the third kind

$$\omega_D(p) = \sum_k \alpha_k \omega_{p_0 p_k}(p), \tag{6.1}$$

where p_0 is an arbitrary point of the surface M which is different from the points of D. The value $\exp \omega_D(p)$ is a multiplicative integral , i.e., a multivalued function possessing a multiplicator (a multiplicative period) in tracing cycles $k_{2\mu}(\mu = 1, \ldots, g)$. The divisor of $\exp \omega_D$ is equal to D. Periods of ω_D along $k_{2\mu}$ are, due to (3.13), equal to

$$\omega_\mu = 2\pi i \sum_k \alpha_k \int_{p_0}^{p_k} dw_\mu \qquad \mu = 1, \ldots, g. \tag{6.2}$$

The analytic function we are interested in is represented by

$$f(p) = \exp\{\omega_D(p) - 2\pi i \sum_{\nu=1}^{g} c_\nu w_\nu(p)\}, \tag{6.3}$$

where c_ν are integers. The function (6.3) is single-valued along the cycles $k_{2\mu-1}(m = 1, \ldots, g)$. In order that the value (6.3) be single-valued along the cycles $k_{2\mu}(\mu = 1, \ldots, g)$, it is necessary and sufficient that there exist such integers c_μ, c'_μ that

$$2\pi i \sum_k \alpha_k w_\mu(p_k) - 2\pi i \sum_{\nu=1}^{g} c_\nu \int_{k_{2\mu}} dw_\nu = 2\pi i c'_\mu, \qquad \mu = 1, \ldots, g.$$

Rewrite these relations, taking into account Eq. (3.7) and (3.7′) in the form

$$\sum_k \alpha_k w_\mu(p_k) = \sum_\nu \left(c_\nu \int_{k_{2\mu}} dw_\nu + c'_\nu \int_{k_{2\mu-1}} dw_\nu \right) = \tag{6.4}$$

$$\sum_\nu \left(c_\nu \int_{k_{2\mu}} dw_\mu + c'_\nu \int_{k_{2\nu-1}} dw_\mu \right), \qquad \mu = 1, \ldots, g.$$

Note that the values $w_\mu(p_k)$ are defined modulo the integer combinations of periods. Therefore, we rewrite (6.4) in the form

$$\sum_k \alpha_k w_\mu(p_k) \equiv 0 \quad (\text{mod periods of} w_\mu), \mu = 1, \ldots, g. \tag{6.5}$$

The system (6.5) is the famous Abel's theorem.

Now consider the boundary problem

$$F^+ = GF^- \quad \text{on} \quad L, \tag{6.6}$$

and suppose that along every connected component of the contour L, the indices of $G(p)$ are equal to zero,

$$\kappa_j = \operatorname{ind}_{L_j} G = \frac{1}{2\pi i}\int_{L_j} d\ln G = 0 \qquad j = 1, \ldots, m.$$

Taking the logarithm of (6.6), we obtain the problem

$$\ln F^+ - \ln F^- = \ln G, \tag{6.7}$$

which was studied in §5. But now solutions can be multivalued with periods multiple of $2\pi i$. Therefore, it is convenient to use some multivalued kernel, for example, (4.3). We obtain the solution in the form

$$\ln F^{\pm}(q) = \frac{1}{2\pi i}\int_L \ln G(p) M^*(p,q)\,dz(p) + 2\pi i \sum_{\nu=1}^{g} c_\nu w_\nu(q), \tag{6.8}$$

where c_μ are integers. Periods of (6.8) along the cycles $k_{2\mu-1}(\mu = 1, \ldots, g)$ are equal to $2\pi i c_\mu$, and therefore single-valuedness conditions for $F^{\pm}(q)$ are

$$\int_L \ln G(p)\,dw_\mu + 2\pi i \sum_\nu c_\nu \int_{k_{2\mu}} dw_\nu = 2\pi i c'_\mu, \qquad \mu = 1, \ldots, g,$$

where c'_μ are integers. As above, these conditions can be rewritten in the form

$$\int_L \ln G\,dw_\mu = 2\pi i \sum_\nu \left\{ -c_\nu \int_{k_{2\mu}} dw_\nu + c'_\nu \int_{k_{2\mu-1}} dw_\nu \right\},$$

i.e.,

$$\int_L \ln G\,dw_\mu \equiv 0 \quad (\text{mod periods of } w_\mu), \qquad \mu = 1, \ldots g. \tag{6.9}$$

Thus, we get the following theorem.

THEOREM 6.1. *In order that the problem (6.6) can be solved, it is necessary and sufficient that (6.9) be valid.*

Abel's theorem can be obtained from (6.9) in the following way (see Rodin [f]). Draw a contour L on the surface M such that all points of the divisor D

belong to the domain D^+ and denote by $D(p)$ an analytic function in T^+ whose poles and zeros are determined by the divisor D.

Then the function $f(q)$ determined by the divisor D, $(f) = D$, is represented by

$$f(q) = D(q)F^+(q) \quad \text{in} \quad T^+$$
$$f(q) = F^-(q) \quad \text{in} \quad T^-, \tag{6.10}$$

where $F^{\pm}(q)$ are analytic functions having no poles and zeros in $T^{\pm}$, respectively. These functions give a solution of the problem

$$F^+(p) = \frac{1}{D(p)} F^-(p) \quad \text{on} \quad L. \tag{6.11}$$

The criterion (6.9) gives

$$\int_L \ln D(p) dw_\mu(p) \equiv 0 \quad (\text{mod periods of } w_\mu). \tag{6.12}$$

The integral on the left-hand side is easily calculated. Connect points $p_k (k = 0, 1, \ldots)$ by cuts (point p_0 is an arbitrary point of the surface different from divisor points) and denote the star which is obtained by s. Then we have

$$\int_L \ln D(p) dw_\mu(p) = -\int_s \ln D(p) dw_\mu(p) = -\sum_k \left\{ \int_{p_0}^{p_k} \ln D(p) dw_\mu \right.$$
$$\left. + \int_{p_k}^{p_0} \ln D(p) dw_\mu \right\} = 2\pi i \sum_k \alpha_k \int_{p_0}^{p_k} dw_\mu$$
$$= 2\pi i \sum_k \alpha_k w_\mu(p_k), \quad \mu = 1, \ldots, g.$$

which involves (6.5).

In conclusion, we note that on the set of divisors of the same degree the equivalency relation can be introduced by

$$D_1 \sim D_2 \text{ if } \deg(D_1 - D_2) = 0,$$
$$D_1 = \sum_j \alpha_j p_j, \qquad D_2 = \sum_k \beta_k q_k \tag{6.13}$$
$$\sum_j \alpha_j w_\mu(p_j) \equiv \sum_k \beta_k w_\mu(q_k) \ (\text{mod periods of } w_\mu), \quad \mu = 1, \ldots, g.$$

The meaning of this equivalency will be clear when we study Jacobi varieties.

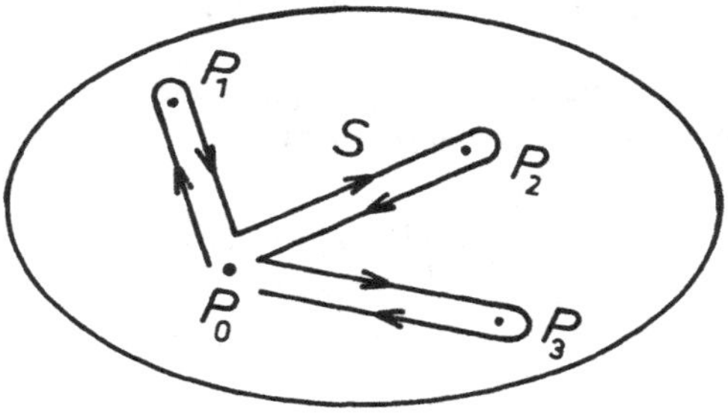

Figure 9

B. *Inversion of Abelian Integrals. The Boundary Problem Solvability*

Now we shall examine an approach to the study of the solvability of the Riemann problem which is different from the methods used in §5. This approach establishes the relation between the Riemann problem and the Jacobi problem of an inversion of Abelian integrals.

Let the contour L consist of $m+1$ connected components, $L = \sum_{j=0}^{m} L_j$. Choose on every L_j a point $p_j (j = 0, \ldots, m)$ and count from these points single-valued branches of the function $\ln G(p)$. It is clear that an increment in $\ln G(p)$, when p passes around L_j, is equal to

$$\kappa_j = \frac{1}{2\pi i} \int_L d \ln G(p), \qquad \sum_j \kappa_j = \kappa. \tag{6.14}$$

Consider the function

$$F_0^{\pm}(q) = \exp \frac{1}{2\pi i} \int_L \ln G(p) M^*(p,q) dz(p), \tag{6.15}$$

where $M^*(p,q)$ is the complex-normalized multivalued kernel (4.3) and branches of $\ln G(p)$ are chosen for use in (6.15). Consider the function $F_0^{\pm}(q)$ on the surface $\widehat{M}$ cut along the cycles of the canonical basis $k_j (j = 1, \ldots, 2g)$ (see §3A) and fix the branch of the kernel by the condition $M^*(p, q_0) = 0$. It is clear that the function (6.15) satisfies the conditon (5.1) and has multiplicative periods along the cycles $k_{2\mu}$,

$$F_\mu = \exp \int_L \ln G dw_\mu(p), \qquad \mu = 1, \dots, g. \tag{6.16}$$

Because $\ln G(p)$ has jumps at $p_j \in L_j$, the Cauchy-type integral has, at these points, logarithmic singularities with principal parts equal to κ_j. Hence, the function $F(q)$ has zeros and poles at points p_k determined by the divisor $(F_0) = \sum \kappa_j p_j, \deg(F_0) = \kappa$. We obtain, therefore, a multiplicative multivalued solution of the problem (5.1).

In order to obtain a single-valued regular solution, it is necessary to choose a multiplicative multivalued function which is a multiple of the divisor $-(F_0)$ and which has periods F^{-1} along $k_{2\mu}(\mu = 1, \dots, g)$. A logarithm of such a function is represented by

$$Z(q) = \sum_{j=0}^{m} \kappa_j \omega_{p_j q_0}(q) + \sum_{k=1}^{\kappa} \omega_{q_0 q_k}(q) - 2\pi i \sum_{\mu=1}^{g} c_\mu w_\mu(q). \tag{6.17}$$

Here ω_{pq} and w_μ are the Abelian integrals (2.13) and (2.16), and c_μ are integers. The points $q_k, k = 1, \dots, \kappa$ are undetermined for the present, and every point p_j repeats κ_j times. The single-valuedness condition for $F(q) \exp Z(q)$ generates the periods relations

$$\begin{aligned}\int_{k_{2\mu}} dZ = 2\pi i \sum_{j=0}^{m} \kappa_j \int_{p_j}^{q_0} dw_\mu + 2\pi i \sum_{k=1}^{\kappa} \int_{q_0}^{q_k} dw_\mu \\ - 2\pi i \sum_{\nu=1}^{g} c_\nu \int_{k_{2\mu}} dw_\nu = 2\pi i c'_\mu - \ln F_\mu, \quad \mu = 1, \dots, g,\end{aligned}$$

where c'_μ are integers. Using Eq. (3.7) and (3.13), rewrite Eq. (6.18) as

$$\sum_{k=1}^{\kappa} w_\mu(q_k) \equiv l_\mu \ (\text{mod periods of } w_\mu)$$

$$l_\mu = -\frac{1}{2\pi i} F_\mu + \sum_{j=0}^{m} \kappa_j w_\mu(p_j), \qquad \mu = 1, \dots, g. \tag{6.19}$$

Therefore, the problem of the existence of a single-valued solution is reduced to the problem of a choice of κ points $q_1, \dots, q_\kappa$ satisfying Eq. (6.19).

THEOREM 6.2. *In order to solve problem (5.1) it is necessary and sufficient that there exist points $q_1, \dots, q_k$ (some points can be repeated) which satisfy the system (6.19).*

The problem (6.19) is known as the Jacobi inversion problem.

THEOREM 6.3. *The Jacobi inversion problem (6.19) is solvable for $\kappa \geqslant g$.*

This statement follows directly from Theorem 5.2. We shall give an independent proof of this important fact.

Let $\delta = q_1 + q_2 + \cdots + q_g$ be a divisor such that $H(\delta) = 0$ and where all points q_j are different. Then the rank of the matrix

$$\Delta = \begin{pmatrix} dw_1(q_1) & \dots & dw_1(q_g) \\ \dots & \dots & \dots \\ dw_g(q_1) & \dots & dw_g(q_g) \end{pmatrix} \tag{6.20}$$

is equal to g. In the opposite case some combinations of rows dw would be multiples of δ that contradict the condition $\dim H(\delta) = 0$. The matrix (6.20) is the Jacobian of the mapping of the g-tuple product $M \times \cdots \times M$ in the g-dimension complex space C^g determined by the formula

$$c_\mu = \sum_{j=1}^{g} w_\mu(q_j), \qquad \mu = 1, \dots, g. \tag{6.21}$$

The relation rank $\Lambda = g$ means that this is a local one-to-one mapping. Hence, if the vector$(c_1, \dots, c_g)$ belongs to some neighbourhood U of zero of C^g, the inversion problem (6.21) can be solved for $\kappa = g$. If $c(c_1, \dots, c_g)$ is an arbitrary vector belonging to C^g, there exists such an integer N that $N^{-1}(c_1, \dots, c_g)$ belongs to U, and hence the inversion problem has a solution $D = \sum s_k, \deg D = g$ for c/N. It is clear that the divisor ND is a solution for the vector c, but $\deg(ND)$ is too large. Consider a divisor $ND - (N-1)gq_0$, where q_0 is a normalized branch point of Abelian integrals of the first kind, $w_\mu(q_0) = 0, \mu = 1, \dots, g$. The degree of this divisor is equal to g and because of the Riemann–Roch theorem,

$$\dim L(ND - (N-1)gq_0 \geqslant 1,$$

and there exists a positive divisor $d = \sum_k \beta_k q_k, \deg d = g$, equivalent to $ND - (N-1)gq_0$. Then, because of Abel's theorem, the divisor

$$D_0 = ND - (N-1)gq_0 - d = \sum \alpha_j p_j, \qquad \deg D_0 = 0,$$

is such that

$$\sum_j \alpha_j w_\mu(p_j) = \sum_j \alpha_j \int_{q_0}^{p_j} dw_\mu \equiv 0 \pmod{\text{periods of } w_\mu} \qquad \mu = 1, \dots, g.$$

This means that

$$\begin{aligned} c_\mu &= N \sum_k w_\mu(s_k) = N \sum_k \int_{q_0}^{s_k} dw_\mu \\ &= \sum_k \beta_k \int_{q_0}^{q_k} dw_\mu \pmod{\text{periods of } w_\mu}, \qquad \mu = 1, \dots, g. \end{aligned}$$

Therefore, the divisor $d, \deg d = g$, is a solution of the inversion problem.

If $\kappa > g$, it is clear that $\kappa - g$ points can be chosen arbitrarily.

The case $\kappa < g$ is more difficult. We shall study this case in §7D.

C. *Jacobi Variety*

Every positive divisor $\delta = \sum_k c_k q_k, \deg \delta = n$ determines the mapping of the n-tuple product of $M, M \times \cdots \times M$, in the g-dimensional complex surface $C^g, l = l(\delta), l(l_1, \ldots, l_g) \in C^g$,

$$l_\mu = \sum_k c_k Z_\mu(q_k), \qquad \mu = 1, \ldots, g, \tag{6.22}$$

where $dZ_\mu (\mu = 1, \ldots, g)$ is the basis of the space of Abelian differentials of the first kind. The value l_μ is determined modulo integer combinations of periods corresponding to integrals Z_μ.

Consider the integer lattice

$$\mathbf{Z}^g = \left\{ \sum_\mu (m_\mu \int\limits_{k_{2\mu-1}} dZ_\nu + n_\mu \int\limits_{k_{2\mu}} dZ_\nu \right\}, \qquad \nu = 1, \ldots, g, \tag{6.23}$$

where m_μ, n_μ are integers. The factor-space $J_g = C^g / \mathbf{Z}^g$ is a g-dimensional torus called *the Jacobi variety of the surface M*. As it follows from Abel's theorem, the divisors d_1 and d_2 are equivalent if and only if the point $l(d_1 - d_2)$ of the Jacobi variety is zero.

On a set of functions $G(p)$ defined on L and such that $\operatorname{ind}_L G(p) = 0$, we define the mapping $s : \{G\} \longrightarrow J_g$ by the relation

$$s(G) \left\{ \int\limits_L \ln G d\omega_\mu, \qquad \mu = 1, \ldots, g \right\}. \tag{6.24}$$

The Riemann problem can be solved in the class of single-valued functions if and only if the point $s(G)$ is zero (Theorem 6.2). If $s(G) \neq 0$, Eq. (6.24) determines multiplicative periods of the solution (6.15).

If $\kappa \neq 0$, the point $s(G)$ also determines a solvability of the Riemann problem. We need Riemann theta-functions to solve this problem.

§7 Riemann Theta-Functions. Solvability of the Riemann Boundary Problem

A. *Zeros of the Riemann Theta-Function*[5]

The theta-function of g variables is defined by the formula

(7.1)
$$\theta(w_1, \dots, w_g) \equiv \theta(w_s) = \sum_{m_j=-\infty}^{\infty} \exp\{\pi i \sum_{k,l=1}^{g} a_{kl} m_k m_l + 2\pi i \sum_{j=1}^{g} m_j w_j\},$$

where the values $a_{kl} = \alpha_{kl} + i\beta_{kl}$ are such that $a_{kl} = a_{lk}$ and the quadratic form

$$\sum \beta_{kl} \xi_k \xi_l \tag{7.2}$$

are positive-definite.

The series (7.1) converges absolutely. In fact,

$$|\exp\{\pi i \sum_{k,l=1}^{g} a_{kl} m_k m_l + 2\pi i \sum_{j=1}^{g} m_j w_j\}| = |\exp\{-\pi \sum_{k,l=1}^{g} \beta_{kl} m_k m_l - 2\pi \sum_{j=1}^{g} m_j \mathrm{Im} w_j\}|.$$

Let K be a maximal value of the quadratic form (7.2) on the unit sphere of C^g. Then,

$$\sum_{k,l=1}^{g} \beta_{kl} m_k m_l \leqslant K \sum_{k=1}^{g} m_k^2. \tag{7.3}$$

Therefore, the series (7.1) converges simultaneously with the series

$$\sum_{m_j=\infty}^{\infty} e^{-\pi K \sum_{k=1}^{g} m_k^2 - 2\pi \sum_{j=1}^{g} m_j \operatorname{Im} w_j} = \prod_{j=1}^{g} \sum_{k=-\infty}^{\infty} \exp\{-\pi K m^2 - 2\pi m \operatorname{Im} w_j\}.$$

A convergence of each series on the right-hand side is established directly.

The following relations

$$\begin{aligned} \theta(w_1 + a_{1s}, w_2 + a_{2s}, \dots, w_g + a_{gs}) &= e^{-i\pi a_{ss} - 2\pi i w_s} \theta(w_1, \dots w_g), \\ \theta(w_1, \dots, w_j + 1, \dots, w_g) &= \theta(w_1, \dots w_g), \qquad s = 1, \dots, g \end{aligned} \tag{7.4}$$

[5] We follow [4].

are valid. The first equality follows from the relation

$$\sum_{k,l=1}^{g} a_{kl}m_k m_l + 2\sum_{j=1}^{g} m_j(w_j + a_{js})$$
$$= \sum_{k,l=1}^{g} a_{kl}(m_k + \delta_{ks})(m_l + \delta_{ls}) + 2\sum_{j=1}^{g}(m_j + \delta_{sj})w_j - a_{ss} - 2w_s,$$

if we take into account that the indices $m_l + 1$ run over all values between $-\infty$ and $+\infty$.

Substituting in (7.1) the Abelian integrals of the first kind $w_j(p)$ (2.13) and assuming that

$$a_{jl} = \int\limits_{k_{2j}} dw_l, \qquad j,l = 1,\ldots,g, \tag{7.5}$$

we obtain the Riemann θ-function $\theta(w_s(p))$ on the Riemann surface $M, p \in M$.

A positive definiteness of the periods matrix $(\operatorname{Im} a_{kl})$ follows from (3.6), and the symmetry $a_{jl} = a_{lj}$ follows from Eq. (3.7).

The function $\theta(w_s(p) - e_s)$, where e_s are arbitrary numbers, is single-valued along cycles $k_{2\mu-1}$ on account of (7.5). The multiplicative periods along cycles k_2 are equal to[6]

$$\tilde{\theta}_\mu = \exp\int\limits_{k_{2\mu}} d\ln\theta$$
$$= \exp\{-\pi i a_{\mu\mu} - 2\pi i w_\mu(p) + 2\pi i e_\mu\}, \quad \mu = 1,\ldots,g. \tag{7.7}$$

THEOREM 7.1. *The function $\theta(w_s(p) - e_s)$ has g zeros on the surface $\widehat{M}$ obtained from M by cutting along $k_j (j = 1,\ldots,2g)$.*

In fact, taking into account that $\partial\widehat{M} = k_1 + k_2 - k_1 - k_2 + \cdots - k_{2g}$, we examine the relation

$$\deg(\theta) = \frac{1}{2\pi i}\int\limits_{\partial\widehat{M}} d\ln\theta(w_s(p) - e_s)$$
$$= \frac{1}{2\pi i}\sum_{\mu=1}^{g}\{-\int\limits_{k_{2\mu-1}} \Delta_{2\mu} d\ln\theta(w_s(p) - e_s)$$
$$+ \int\limits_{k_{2\mu}} \Delta_{2\mu-1} d\ln\theta(w_s(p) - e_s)\}.$$

---

[6]Here and below the values of multivalued functions on the cycles k_j are taken at points $p \in k_j^+$ (see Figure 3).

Here the symbol $\Delta_j f$ means an increment of a function f obtained in tracing of the cycle k_j. With regard to (7.7) and (2.13), we have

$$\deg\theta = -\frac{1}{2\pi i}\sum_{\mu=1}^{g}\int_{k_{2\mu-1}} d\ln\tilde{\theta}_\mu(p) = \sum_{\mu=1}^{g}\int_{k_{2\mu-1}} dw_\mu = g. \tag{7.8}$$

Denote the zeros of $\theta(w_s(p)-e_s)$ by $p_1,\ldots,p_2$, $(\theta)=\sum_{j=1}^{g} p_j$.

B. *The Problem of Inversion of Abelian Integrals*

Calculate the sum

$$\begin{aligned} l_\mu(\theta) &= \sum_{j=1}^{g} w_\mu(p_j) = \frac{1}{2\pi i}\int_{\partial\widehat{M}} w_\mu(p)d\ln\theta(w_s(p)-e_s)\\ &= \frac{1}{2\pi i}\sum_{\nu=1}^{g}\Big\{-\int_{k_{2\nu-1}}\Delta_{2\nu}\left[w_\mu(p)d\ln\theta(w_s(p)-e_s)\right]\\ &\quad+\int_{k_{2\nu}}\Delta_{2\nu-1}\left[w_\mu(p)d\ln\theta(w_s(p)-e_s)\right]\Big\},\\ \mu &= 1,\ldots,g. \end{aligned}\tag{7.9}$$

By (7.5) and (7.7), we have

$$\Delta_{2\nu}w_\mu(p)=a_{\mu\nu},\qquad \Delta_{2\nu}d\ln\theta(w_s(p)-e_s)=-2\pi i dw_\nu(p).$$

Further

$$\begin{aligned}\Delta_{2\nu}\{w_\mu(p)d\ln\theta(w_s(p)-e_s)\} &= [w_\mu(p)+a_{\mu\nu}][d\ln\theta(w_s(p)\\ &\quad -e_s)-2\pi i dw_\nu(p)]-w_\mu(p)d\ln\theta(w_s(p)-e_s)\\ &= a_{\mu\nu}d\ln\theta(w_s(p)-es)-2\pi i[w_\mu(p)+a_{\mu\nu}]dw_\nu.\end{aligned}$$

Therefore, the first integral on the right-hand side of Eq. (7.9) is equal to

$$\begin{aligned} -\frac{a_{\mu\nu}}{2\pi i}\int_{k_{2\nu-1}} d\ln\theta(w_s(p)-es)+\int_{k_{2\nu-1}} w_\nu dw_\nu + a_{\mu\nu}\int_{k_{2\nu-1}} dw_\mu\\ = -a_{\mu\nu}n_\nu+\int_{k_{2\nu-1}} w_\mu dw_\nu + a_{\mu\nu},\qquad \mu=1,\ldots,g,\end{aligned}\tag{7.10}$$

where n_ν are integers. Further,

$$\Delta_{2\nu-1}w_\mu(p)=\delta_{\mu\nu},\qquad \Delta_{2\nu-1}d\ln\theta(w_s(p)-es)=0.$$

The second integral on the right-hand side of Eq. (7.9) is equal to

$$\frac{\delta_{\mu\nu}}{2\pi i}\int_{k_{2\mu}} d\ln\theta(w_s(p)-e_s)=\delta_{\mu\nu}\left(-\frac{a_{\mu\mu}}{2}-w_\mu(p)+e_\mu\right).$$

Thus, we obtain

$$\sum_{j=1}^{g} w_\mu(p_j) = \sum_{\nu=1}^{g}(-a_{\mu\nu}n_\nu + a_{\mu\nu} + \int_{k_{2\nu-1}} w_\mu dw_\nu) - \frac{a_{\mu\mu}}{2} - w_\mu(p) + e_\mu.$$

One integral on the right-hand side is calculated by parts

$$\int_{k_{2\mu-1}} w_\mu dw_\mu = \Delta_{2\mu-1}(w_\mu)^2 - \int_{k_{2\mu-1}} w_\mu dw_\mu,$$

whence

$$\int_{k_{2\mu-1}} w_\mu dw_\mu = w_\mu(p) + \frac{1}{2}. \tag{7.11}$$

We obtain

(7.12)

$$\sum_{j=1}^{g} w_\mu(p_j) = \sum_{\nu=1}^{g}(-a_{\mu\nu}n_\nu + a_{\mu\nu}) + \sum_{\nu\neq\mu}^{1,\dots,g} \int_{k_{2\nu-1}} w_\mu dw_\nu$$
$$+ w_\mu(p) + \frac{1}{2} - \frac{a_{\mu\mu}}{2} - w_\mu(p) + e_\mu \equiv e_\mu - k_\mu \quad (\text{mod periods of } w_\mu)$$

where

$$k_\mu = -\frac{1}{2} + \frac{a_{\mu\mu}}{2} - \sum_{\nu\neq\mu}^{1,\dots,g} \int_{k_{2\nu-1}} w_\mu dw_\nu \tag{7.12$'$}$$

are so-called Riemann constants which depend only on the surface and not on the divisor.

Therefore, we have proved the following theorem.

THEOREM 7.2. *The solution of the inversion problem*

$$\sum_{j=1}^{g} w_\mu(p_j) \equiv e_\mu - k_\mu \quad (\text{mod } \textit{periods of} w_\mu) \tag{7.13}$$

is a divisor of zeros of the function $\theta(w_s(p) - e_s)$.

Note that for a set of parameters $\{e_s\}$, the function $\theta(w_s(p) - e_s)$ is identically equal to zero, and Theorem 7.2 loses its meaning.

In that case it can be verified that the functions

$$\frac{\partial\theta(w_s(p) - e_s)}{\partial w_j}, \qquad j = 1, \dots, g \tag{7.14}$$

have the same periods as $\theta(w_s(p) - e_s)$. This means that every nonzero function (7.14) has g zeros on $\widehat{M}$, and these zeros give the solution of the inversion problem (7.13). If all first derivatives are identically equal to zero, the second derivatives can be used, and so on.

C. *Divisor Classes*

THEOREM 7.3. *Let $D = \sum_{j=1}^{g} p_j \geqslant 0$ be an arbitrary divisor (some points p_j may coincide),* $\deg D = g$, *and*

$$e_\mu = \sum_{j=1}^{g} w_\mu(p_j) + k_\mu, \qquad \mu = 1, \dots, g. \tag{7.15}$$

Then $\theta(w_s(p) - e_s)$ has zeros at the points of D.

First, consider the case $\dim L(D) = 1$. If the θ-function does not vanish identically and $p'_1, \dots, p'_g$ are its zeros, then

$$\sum_{j=1}^{g} w_\mu(p_j) \equiv \sum_{l=1}^{g} w_\mu(p'_l) \qquad (\text{mod periods of } w_\mu), \qquad \mu = 1, \dots, g. \tag{7.16}$$

This involves an equivalence of divisors $\sum p_j$ and $\sum p'_l$ and means that these divisors coincide, since $\dim L(D) = 1$.

If $\dim L(D) > 1$, then due to the Riemann–Roch theorem,

$$\dim H(D) = \dim L(D) - 1 > 0.$$

Therefore, there exist Abelian differentials which are multiples of D. Fix points $p_1, \dots, p_r, r = g - \dim H(D) - 1$, and shift the points $p_{r+1}, \dots, p_g$ by some small amount. A new divisor is obtained which is denoted by $D' = \sum p'_k$. It is clear that there exists no Abelian differential which would be multiples of $D', \dim H(D') = 0$.

Denoting

$$e'_\mu = \sum_j w_\mu(p'_j) + k_\mu, \qquad \mu = 1, \dots, g, \tag{7.17}$$

we obtain that $\theta(w_s(p'_j) - e_s) = 0$. Proceeding to the limit as $p'_j \to p_j$, we conclude that $\theta(w_s(p_j) - e_s) = 0, p_j \in D$.

THEOREM 7.4. *If $\dim L(D) > 1, deg D = g$ (such divisors are called special), then*

$$\theta(w_s(p) - \sum_{j=1}^{g} w_s(p_j) - k_s) \equiv 0, \qquad p \in M.$$

In fact, due to Theorem 7.3, this function vanishes at every divisor of the equivalency class. On the other hand, however, it has only g zeros, and the statement in the theorem follows.

THEOREM 7.5. *For any $g-1$ points $p_1, \dots, p_{g-1} \in M$,*

$$\theta(\sum_{j=1}^{g-1} w_s(p_j) + k_s) = 0. \tag{7.18}$$

To see this, let us examine the Abelian differential of the first kind vanishing at $p_1, \dots, p_{g-1}$. Denote by p_g some zero of this differential which is different from $p_1, \dots, p_{g-1}$ and assume that $D = \sum_1^g p_j$. Then $H(D) > 1$. Therefore,

$$\theta(w_s(p) - \sum_{j=1}^{g} w_s(p_j) - k_s) = 0 \qquad \text{for all} \quad p \in M.$$

Supposing that $p = p_g$ and taking into account that $\theta(w_1, \dots, w_g)$ is an even function, we obtain Eq. (7.18).

The inverse theorem is valid also.

THEOREM 7.6. *If*

$$\theta(w_s(p) - e_s) = 0 \qquad \textit{for all} \quad p \in M,$$

then there exist points $p_1, \dots, p_g$ *such that*

$$e_s \equiv \sum_{j=1}^{g} w_s(p_j) + k_s \quad (\text{mod } \textit{periods of } w_s), \tag{7.19}$$

and $\dim(\sum_j^g p_j > 1$ *(see [4], [7]).*

D. *The Solvability of the Riemann Problem*

We shall consider conditions of solvability of the Riemann boundary problem in the case $0 < \kappa < g$.

As was shown in §6B, this problem is reduced to the inversion problem (6.19)

$$\sum_{j=1}^{\kappa} w_\mu(q_j) \equiv l_\mu \quad (\text{mod periods of } w_\mu). \tag{7.20}$$

THEOREM 7.7. *In order that the inverse problem (7.20) can be solved, it is necessary and sufficient that for any system of points* $\tilde{q}_1, \dots, \tilde{q}_{g-\kappa-1}$ *the equation*

$$\theta(\sum_{j=1}^{g-\kappa-1} w_s(\tilde{q}_j) + l_s + k_s) = 0 \tag{7.21}$$

be valid.

Let problem (7.20) be solvable, $q_1, \dots q_\kappa$ be a solution, and $\tilde{q}_1, \dots, \tilde{q}_{g-\kappa-1}$ be arbitrary points. Then the set of $g-1$ points $q_1, \dots, q_\kappa, \tilde{q}_1, \dots, \tilde{q}_{g-\kappa-1}$ possesses the property

$$\theta(\sum_{j=1}^{\kappa} w_s(q_j) + \sum_{i=1}^{g-\kappa-1} w_s(\tilde{q}_i) + k_s) = 0 \tag{7.22}$$

because of Eq. (7.18). Substituting values l_s from Eq. (7.20) into the second sum in Eq. (7.22), we obtain (7.21).

Conversely, let values $l_s (s = 1, \dots, g)$ be such that Eq. (7.21) is valid for any points $\tilde{q}_1, \dots, \tilde{q}_{g-\kappa-1}$. Rewrite (7.21) in the form

$$\theta(w_s(p) + \sum_{j=1}^{g-\kappa-1} w_s(\tilde{q}_j) + l_s + k_s - w_s(p)) = 0.$$

The function θ is even, and hence

$$\theta(w_s(p) - (w_s(p) + \sum_{j=1}^{g-\kappa-1} w_s(\tilde{q}_j) + l_s + k_s)) = 0.$$

Denote

$$e_s = w_s(p) + \sum_{j=1}^{g-\kappa-1} w_s(\tilde{q}_j) + l_s + k_s.$$

From Theorem 7.6, there exist some points $p_1, \dots, p_g$, such that

$$w_s(p) + \sum_{j=1}^{g-\kappa-1} w_s(\tilde{q}_j) + l_s + k_s = \sum_{j=1}^{g} w_s(p_j) + k_s \quad (\text{mod periods of } w_s).$$

Now we fix the arbitrary points $p, \tilde{q}_j$,

$$p = p_g, \qquad \tilde{q}_j = p_{\kappa+j} \qquad j = 1, \dots, g - \kappa - 1.$$

We obtain

$$\sum_{j=1}^{\kappa} w_s(p_j) \equiv l_s \quad (\text{mod periods of } w_s).$$

§8 Explicit Formulae for Solutions of the Riemann Problem

Let $f(q)$ be some function with multiplicative periods

$$f_{2\nu} = \exp \int_{k_{2\nu}} d \ln f, \qquad \nu = 1, 2, \dots, g$$

along the cycles $k_{2\nu}$, and 1 along $k_{2\nu-1} (\nu = 1, \dots, g)$. The function $f(q)$ satisfies the boundary condition (see Figure 3),

$$f^+(p) = \frac{1}{f_{2\nu}} f^-(p), \qquad p \in k_{2\nu-1}. \tag{8.1}$$

As was noted by Zverovich [41], the function $\theta(w_s(p) - e_s)$ is a solution of the Riemann problem (see (7.7)),

$$\begin{aligned} \theta^+(p) &= \frac{1}{\tilde{\theta}(p)}\theta^-(p), \qquad p \in K = \bigcup_{\nu=1}^{g} k_{2\nu-1}, \\ \tilde{\theta}(p) &= \exp\{-\pi i - 2\pi i w_\nu(p) + 2\pi i e_\nu\}, \qquad p \in k_{2\nu-1}. \end{aligned} \tag{8.2}$$

We use this observation to obtain the explicit formulae for solving the Riemann problem. Below we assume, for simplicity, that the cycles $k_{2\nu-1}$ and the contour L have zero intersection.

Case $\kappa = g$. Since the solution of the problem (8.2) is the θ-function having g zeros, the index of this problem is equal to g. This follows also from the relation

$$\frac{1}{2\pi i}\int_K d\ln\tilde{\theta}^{-1}(p) = \sum_{\nu-1}^{g} dw_\nu = g. \tag{8.3}$$

Consider the boundary problem

$$\begin{aligned} F^+(p) &= G(p)F^-(p), \qquad p \in L = \bigcup_{j=1}^{m} L_j, \\ \kappa &= \operatorname{ind}_L G = \sum_{j=1}^{m} \operatorname{ind}_{L_j} G = \sum_{j=1}^{m} \kappa_j = g. \end{aligned} \tag{8.4}$$

Let the values l_s be determined by Eq. (6.19) and

$$e_s = k_s + l_s, \qquad s = 1, \dots, g, \tag{8.5}$$

where k_s are the Riemann constants (7.12′). First, let us suppose that $\theta(w_s(p) - e_s) \not\equiv 0$. Then the zeros of $\theta(w_s(p) - e_s)$ coincide with zeros of the solution of the problem (8.4) (see (6.19) and Theorem 7.2). Thus, the boundary problem[7]

$$\begin{aligned} \Phi^+(p) &= \tilde{G}(p)\Phi^-(p), \qquad p \in \tilde{L} = L \cup K, \\ \tilde{G}(p) &= \begin{cases} G(p), & p \in L \\ \tilde{\theta}(p), & p \in K \end{cases} \end{aligned} \tag{8.6}$$

has the single regular solution

$$\Phi(p) = F(q)\theta^{-1}(w_s(q) - e_s). \tag{8.7}$$

---

[7] If $L = K, \tilde{G}(p) = G(p)\tilde{\theta}(p)$.

Fix arbitrary points $q_k \in L_k (k = 1, \ldots, m), p_j \in k_{2j-1} (j = 1, \ldots, g), s_0 \in M$. We count off the branches of the function $\ln G(p)$ from these points. Then

$$\begin{aligned}(8.8) \qquad \Phi(q) = \exp \Bigg\{ &\frac{1}{2\pi i} \int_L \ln G(p) M^*(p,q) dz(p) \\ &+ \sum_{l=1}^{g} \int_{k_{2l-1}} (-\frac{1}{2} a_{ll} - w_l(p) - e_l) M^*(p,q) dz(p) \\ &- \sum_{j=1}^{m} \kappa_j \omega_{s_0 q_j}(q) + \sum_{k=1}^{g} \omega_{s_0 p_k}(q) \Bigg\}.\end{aligned}$$

This function is holomorphic in $M \backslash L$, different from zero, and satisfies the boundary condition (8.6). We want to verify that the function

$$(8.9) \qquad F(q) = \Phi(q) \theta(w_s(p) - e_s)$$

is single-valued. We shall calculate its periods along the cycles $k_{2\nu} (\nu = 1, \ldots, g)$.

As is obvious from Figure 2 and 3, the periods of a function $f(q)$ along the cycles k_{2j} are equal to the jumps of the function $f(q)$ on the cuts k_{2j-1} (see Figure 7),

$$(8.10) \qquad \Delta_{2j} f = \int_{k_{2j}} df = f-(p) - f+(p), \qquad q \in k_{2j-1}, \qquad j = 1, \ldots, g.$$

Consider the periods of the integral

$$(8.11) \qquad f(q) = \frac{1}{2\pi i} \int_{k_{2j-1}} \varphi(p) M^*(p,q) dz(p)$$

along the cycle k_{2j}. We have (see (4.7))

$$\begin{aligned} f^+(q) &= \frac{1}{2}\varphi(q) + \frac{1}{2\pi i} \int_{k_{2j-1}} \varphi(p) M^*(p,q) dz(p), \qquad q \in k_{2j-1}^+, \\ (8.12) \qquad f^-(q) &= -\frac{1}{2}\varphi(q) + \frac{1}{2\pi i} \int_{k_{2j-1}} \varphi(p) [M^*(p,q) + w_j'(p) \cdot 2\pi i] dz(p), \\ \Delta_{2j} f &= \int_{k_{2j}} df = -\varphi(q) + \int_{k_{2j-1}} \varphi(p) dw_j(p), \qquad j = 1, \ldots, g. \end{aligned}$$

Calculate now the multiplicative periods of the function (8.9)

$$F_{2j} = \exp \int_{k_{2j}} d \ln F, \qquad j = 1, \ldots, g.$$

Let

$$\delta_j(q) = \begin{cases} 0 & q \in k_{2j-1}, \\ 1 & q \in k_{2j-1}, \end{cases} \qquad j = 1, \ldots, g.$$

Taking into account (4.13), (8.12), (3.13), and (7.7), we obtain

$$\begin{aligned} F_{2j} = \exp\Bigg\{ & \int_L \ln G dw_j + 2\pi i \sum_{l=1}^{g} \int_{k_{2l-1}} (-\frac{1}{2}a_{ll} - w_l(p) \\ & + e_l) dw_j(p) - 2\pi i \sum_{l=1}^{g} (-\frac{1}{2}a_{ll} - w_l(q) + e_l)\delta_l(q) - 2\pi i \sum_{s=1}^{m} \kappa_s w_j(q_s) \\ & + 2\pi i \sum_{k=1}^{g} w_j(p_k) + \sum_{l=1}^{g} [-\pi i a_{ll} - 2\pi i w_l(q) + 2\pi i e_l]\delta_l(q) \Bigg\} \\ = \exp\Bigg\{ & \int_L \ln G dw_j + 2\pi i \sum_{l=1}^{g} \int_{k_{2l-1}} (-\frac{1}{2}a_{ll} - w_l(p) + e_l) dw_j(p) \\ & - 2\pi i \sum_{s=1}^{m} \kappa_s w_j(q_s) + 2\pi i \sum_{k=1}^{g} w_j(p_k) \Bigg\}. \end{aligned}$$

Because of (2.13),

$$\begin{aligned} F_{2j} = \exp\Bigg\{ & \int_L \ln G dw_j - 2\pi i (\frac{1}{2}a_{jj} - e_j) \\ & - 2\pi i \sum_{l=1}^{g} \int_{k_{2l-1}} w_l dw_j - 2\pi i \sum_{s=1}^{m} \kappa_s w_j(q_s) + 2\pi i \sum_{k=1}^{g} w_j(p_k) \Bigg\}. \end{aligned}$$

Take into account (8.5), (6.16), and (6.19). Then

$$\begin{aligned} F_{2j} = \exp\{ & \int_L \ln G dw_j - 2\pi i (\frac{1}{2}a_{jj} - k_j) + \sum_{s=1}^{m} \kappa_s w_j(q_s) \\ & - \int_L \ln G dw_j - 2\pi i \sum_{l=1}^{g} \int_{k_{2l-1}} w_l dw_j - 2\pi i \sum_{s=1}^{m} \kappa_s w_j(q_s) + 2\pi i \sum_{k=1}^{g} w_j(p_k) \\ = \exp\{ & -2\pi i (\frac{1}{2}a_{jj} - k_j) - 2\pi i \sum_{l=1}^{g} \int_{k_{2l-1}} w_l dw_j + 2\pi i \sum_{k=1}^{g} w_j(p_k) \}. \end{aligned}$$

By (7.11) and (7.12′),

$$(8.13)\qquad F_{2j} = \exp\Bigg\{-2\pi i\Big(\frac{1}{2}a_{jj} + \frac{1}{2} - \frac{1}{2}a_{jj} + \sum_{l\neq j}^{1,\dots,g}\int\limits_{k_{2l-1}} w_j dw_l\Big)$$

$$- 2\pi i\sum_{l\neq j}^{1,\dots,g}\int\limits_{k_{2l-1}} w_l dw_j - 2\pi i\Big(w_j(p_j) + \frac{1}{2}\Big)$$

$$+2\pi i\sum_{k=1}^{g} w_j(p_k)\Bigg\}$$

$$= \exp\Bigg\{-2\pi i\sum_{r\neq j}^{1,\dots,g}\int\limits_{k_{2l-1}} d(w_j w_r) - w_j(p_k)\Bigg\}$$

$$= \exp\Bigg\{-2\pi i\sum_{r\neq j}^{1,\dots,g}[(w_r(p_r)+1)w_j(p_r)$$

$$-w_j(p_r)w_r(p_r) - w_j(p_r)]\Bigg\} = 1.$$

Therefore, the function (8.9) is a solution of the boundary problem.

If $\theta(w_s(q) - e_s)$ is identically equal to zero, it is necessary to use a function θ_α, as was pointed out above (see (7.14)).

We have

(8.14)

$$F^{\pm}(q) = \theta_\alpha(w_s(q) - k_s - l_s)\exp\Bigg\{\frac{1}{2\pi i}\int\limits_{L}\ln G(p)M^*(p,q)dz(p)$$

$$+\sum_{r=1}^{g}\int\limits_{k_{2r-1}}\Big(-\frac{1}{2}a_{rr} - w_r(p) - l_r - k_r\Big)M^*(p,q)dz(p)$$

$$+\sum_{j=1}^{m}\kappa_j\omega_{s_0 q_j}(q) - \sum_{k=1}^{g}\omega_{s_0 p_k}(q)\Bigg\}.$$

Note that the value

$$\int\limits_{k_{2j-1}} M^*(p,q)dz(p)$$

is equal to zero. Indeed, by (4.3), (4.1), and (2.16),

$$\int\limits_{k_{2j-1}} M^*(p,q)dz(p) = \int\limits_{k_{2j-1}} d_p[\omega_{p_0 p}(q) - \omega_{p_0 p}(q)]$$

$$= \int\limits_{k_{2j-1}} d_p[\omega_{q_0 q}(p) - \omega_{q_0 q}(p_0)] = \int\limits_{k_{2j-1}} d\omega_{q_0 q}(p) = 0.$$

Therefore, the formula (8.14) takes the form

$$F^{\pm}(q) = \theta_\alpha(w_s(q) - k_s - l_s) \exp\left\{ \frac{1}{2\pi i} \int_L \ln G(p) M^*(p,q)\, dz(p) \right. \tag{8.15}$$
$$\left. + \sum_{r=1}^{g} \int_{k_{2r-1}} w_r(p) M^*(p,q)\, dz(p) + \sum_{j=1}^{m} \kappa_j \omega_{s_0 q_j}(q) - \sum_{k=1}^{g} \omega_{s_0 p_k}(q) \right\}.$$

The case $\kappa > g$. Fix $\kappa - g$ arbitrary points $s_1, \dots, s_r \in T^+$, $r \leqslant \kappa - g$, $s_{r+1}, \dots, s_{\kappa-g} \in T^-$. Define the analytic functions $\gamma^{\pm}(q)$ in T, such that $(\gamma^+) = \sum_1^r s_j, (\gamma^-) = \sum_{r+1}^{\kappa-g} s_j$. Then we look for the solution of the problem (8.4) in the form of

$$F^{\pm}(q) = \gamma^{\pm}(q) F_0^{\pm}(q), \qquad q \in T^{\pm}. \tag{8.16}$$

We obtain the problem

$$F_0^+(p) = G(p) \frac{\gamma^-(p)}{\gamma^+(p)} F_0^-(p) \tag{8.17}$$

of the index g. Therefore, in this case the solution depends on an arbitrary divisor of the degree $\kappa - g$.

The case $0 \leqslant \kappa < g$. Let the divisor $\gamma^+ \in T^+, \deg \gamma^+ = g - \kappa$. Assume

$$F^+(q) = \frac{F_0^+(q)}{\gamma^+(q)}, \qquad F^-(q) = F_0^-(q). \tag{8.18}$$

We obtain the problem

$$F_0^+(p) = G(p) \gamma^+(p) F_0^-(p) \tag{8.19}$$

of index g. The functions $F_0^{\pm}$ are determined by formula (8.14). Formula (8.18) determines a holomorphic solution if the function (8.18) is a multiple of the divisor γ^+.

CHAPTER 2

COMPLEX VECTOR BUNDLES OVER COMPACT RIEMANN SURFACES

§9 De Rham and Dolbeault Theorems

1. Define for every open set of points U on a closed Riemann surface a module (group, vector space) $H(U)$, so that for each pair of open sets $U \in V$ there is a homorphism

$$\psi_U^V : H(V) \to H(U) \tag{9.1}$$

satisfying the condition

$$\psi_U^V \psi_V^W = \psi_U^W, \qquad U \subset V \subset W.$$

This homomorphism is called a *restriction of* $H(V)$ *on* U. The set $P = \{H(U), \psi_U^V\}$ is called a *projective system of modules* (groups, vector spaces) or a *presheaf over the surface* M. Elements of $H(U)$ are called *sections of the presheaf* P *over* U; the group of sections is denoted by $\Gamma(P, U)$.

Let two presheaves $P = \{H(U), \psi_U^V\}$ and $P' = \{H'(U), \psi_U'^V\}$ over M be given. A system of homomorphisms $r = \{r_U\}$, defined for every open set U

$$r_U : H(U) \to H'(U)$$

satisfying a commutativity condition $\psi'^V_U r_V = r_U \psi_U^V$, is called a *homomorphism of presheaves*, $r : P \to P'$. In particular, if all r_U are monomorphisms, we obtain an embedding of presheaves. The quotient-presheaf P/P' is defined by the quotient-modules $H(U)/H'(U)$.

The homomorphism sequence

$$\cdots \to P' \xrightarrow{\alpha} P \xrightarrow{\beta} P'' \to \ldots$$

is called exact if $\beta \circ \alpha = 0$, and $\operatorname{Im} \alpha = \operatorname{Ker} \beta$. In particular, for any pair of presheaves $P' \subset P$ one can construct the exact sequence

$$0 \to P' \xrightarrow{i} P \xrightarrow{\pi} P'' \to 0, \tag{9.2}$$

where i is an embedding, π is a projection on the quotient-presheaf $P'' = P/P'$. Such sequences are called *short*.

A presheaf P is called a sheaf if two conditions are satisfied.

a) Let $\{U_i\}$, $i \in I$, be a family of open sets on M, U be the union of these sets, and $s_1, s_2 \in H(U)$. Then, if the restrictions of s_1 and s_2 on each set U_i belonging to U_i coincide, it is necessary that $s_1 = s_2$.

b) Let $s_i \in H(U_i)$ be such that for any $i, j \in I$ the restrictions of s_i and s_j are equal on $U_i \cap U_j$. Then there exists $s \in H(U)$ whose restriction on U_i is equal to s_i for any $i \in I$. These notions are treated in many books; for instance, [5], [7], [16], [17], and [20]. Let us list some examples of sheaves as used below.

Differential forms $\omega = a dz$ and $\omega = a d\bar{z}$ are called forms of type $(1,0)$ and $(0,1)$, respectively. Differential forms $\omega = a dz \wedge d\bar{z}$ are called forms of type $(1,1)$. Forms of type $(0,0)$ (or 0) are functions on M.

Let $H(U)$ be a linear space of forms of type (α, β) of the class C^∞ in the domain U. The corresponding sheaf is called a *sheaf of germs of differential forms* of type $(\alpha,\beta)(\alpha,\beta = 0,1)$ and is denoted by $A^{\alpha,\beta}$. We shall also consider subsheaves of these sheaves. $C^{\alpha,\beta}$ is a sheaf of germs of closed differential forms of type (α, β) (the definition of the closed linear forms was given in §2; a closed form of type 0 is a constant; a closed form of type (1,1) is any form).[1] $Q^{\alpha,\beta}$ is a sheaf of germs of holomorphic forms of the type (α, β). We write also Q in place of $Q^{0,0}$.

Another example of a sheaf is given by the multiplicative groups $H(U)$ of functions meromorphic in U. We denote this sheaf by M^*. Its subsheaf formed by germs of holomorphic functions different from zero is denoted by Q^*.

2. Let $\{U_i, i \in I\}$ be some covering of the surface M by simply-connected domains. Every finite set of indices $i_0, \ldots, i_q$ determines the domain $U_{i_0,\ldots,i_q} = U_{i_0} \cap \cdots \cap U_{i_q}$. Consider a sheaf $P = \{H(U), \psi_U^V\}$ and associate with every domain $U_{i_0,\ldots,i_q}$ some element $s_{i_0,\ldots,i_q} \in H_{i_0,\ldots,i_q}$. This correspondence defines a q-cochain $s_{i_0,\ldots,i_q}$ with values in P. The group $Ch^q(P,\{U_i\})$ of q-cochains with values in P corresponding to a covering $\{U_i\}$ is defined as a group of finite linear sums $\sum_k \alpha_k s_k^q$, where s_k^q are q-cochains. We define the coboundary operator

$$\delta : Cn^q(P,\{U_i\}) \to Cn^{q+1}(P,\{U_i\})$$

by the formula $(c = \{s_{i_1}, \ldots, i_q\})$,

$$(\delta c)_{i_0,\ldots,i_{q+1}} = \sum_{j=0}^{q+1} (-1)^j \psi_{U_{i_0,\ldots,i_q}}^{U_{i_0,\ldots,\hat{i}_j,\ldots,i_{q+1}}} s_{i_0,\ldots,\hat{i}_j,\ldots,i_{q+1}}. \tag{9.3}$$

A cochain c with zero coboundary, $\delta c = 0$, is called a cocycle. The groups of q-cocycles are denoted by $Z^q(P,\{U_i\})$. A q-cocycle c that is a coboundary of

[1] Therefore for the closed form ω of class C^1, $d\omega = 0$.

some $(q-1)$-cochain c', $c = \delta c'$ is said to be cohomological to zero. Such cocycles form the group $B^q(P,\{U_i\}) = \delta Cn^{q-1}(P,\{U_i\}), q > 0, B^0(P,\{U_i\}) \stackrel{\text{def}}{=} 0$.

Groups of cohomologies of a covering $\{U_i\}$ with coefficients in sheaf P are defined by the relation

$$H^q(P,\{U_i\}) = Z^q(P,\{U_i\})/B^q(P,\{U_i\}). \tag{9.4}$$

By refining coverings, it is possible to proceed to the projective limits called the *cohomology groups of the surface.* But we use only the groups (9.4). The covering is fixed, and the symbol $\{U_i\}$ is omitted below. We shall write $H^q(P)$ or $H^q(P,M)$.

Consider a 0-cochain $\{s_i\}$ where functions s_i are defined in the domains U_i. A coboundary of this cochain $\delta\{s_i\} = \{s_{ij}\}$ is the 1-cochain defined on the intersections of domains

$$s_{ij} = s_i - s_j \qquad \text{on} \quad U_i \cap U_j.$$

Therefore, if $\{s_i\}$ is a cocycle, $\delta(s_i) = 0$, then $s_i = s_j$ in all intersections $U_i \cap U_j$, and hence $\{s_i\}$ forms a section over M. Thus, we obtain an important relation,

$$H^0(P,M) = \Gamma(P,M), \tag{9.5}$$

where $\Gamma(P,M)$ is a group of sections of a sheaf over M. For the sheaf $A^{\alpha,\beta}$ it is the group of smooth (α,β)-forms on the surface M.

To every short exact sequence of sheaves (9.2), there is a corresponding exact cohomology sequence

$$\begin{aligned} 0 \longrightarrow H^0(P') \xrightarrow{i^0} H^0(P) \xrightarrow{\pi^0} H^0(P'') \xrightarrow{\delta_0^*} H^1(P') \xrightarrow{i^1} H^1(P) \\ \xrightarrow{\pi^1} H^1(P'') \xrightarrow{\delta_1^*} \ldots \longrightarrow H^q(P') \xrightarrow{i^q} H^q(P) \xrightarrow{\pi^q} H^q(P'') \xrightarrow{\delta_q^*} \ldots. \end{aligned} \tag{9.6}$$

The induced homomorphisms i^q, π^q conform to homomorphisms of the corresponding groups of sections and are defined on cohomology groups since they commute with the coboundary operator. We can define the homomorphisms δ_q^*. Let a cocycle $\{s_{i_0,\ldots,i_q}\}$ represent some element of the cohomology group $H^q(P'')$. If $P'' = (M''(U), \varphi_U^V)$, then the value $s_{i_0}, \ldots, i_q \in M''(U_{i_0,\ldots,i_q})$ and is an image of some class of elements belonging to the group $M(U_{i_0,\ldots,i_q})$ (see (9.2)). If $s'_{i_0,\ldots,i_q}, s''_{i_0,\ldots,i_q} \in M(U_{i_0,\ldots,i_q})$ are two elements of this class being a preimage of $s_{i_0,\ldots,i_q}$, then $s'_{i_0,\ldots,i_q} - s''_{i_0,\ldots,i_q} = s^0_{i_0,\ldots,i_q} \in M'(U_{i_0,\ldots,i_q})$. Thus we have obtained the cochain $\{s^0_{i_0,\ldots,i_q}\} \in Cn^q(P')$. We define

$$\delta_q^* \{s_{i_0,\ldots,i_q}\} = \delta\left\{s^0_{i_0,\ldots,i_q}\right\} \in Cn^{q+1}(P').$$

Consider the short exact sequence

$$0 \longrightarrow C \xrightarrow{i} A^0 \xrightarrow{d} C^1 \longrightarrow 0, \tag{9.7}$$

where C^1 is the sheaf of germs of closed differentials of the type $\omega = adz + bd\bar{z}$, and C is a constant sheaf. The exactness of the sequence (9.7) follows from the fact that in a simply-connected domain every closed 1-form is exact, and consequently the operator $d = \frac{\partial}{\partial z(p)} \wedge dz(p) + \frac{\partial}{\partial \overline{z(p)}} \wedge dz(p)$ is an epimorphism on the space of smooth functions.

From (9.6) we have

$$0 \to H^0(C,M) \xrightarrow{i^0} H^0(A^0,M) \xrightarrow{d^0} H^0(C^1,M)$$
$$\xrightarrow{\delta_0^*} H^1(C,M) \xrightarrow{i^1} H^1(A^0,M) \xrightarrow{d^1} H^1(C^1,M) \xrightarrow{\delta_1^*} \ldots.$$

Consider a partition of unity subordinated to the covering $\{U_i\}$ fixed above, i. e., the set of infinitely differentiable functions $\alpha_i(p)$ whose supports belong to the domains U_i, such that for every $p \in M$, $\sum_i \alpha_i(p) \equiv 1$. The existence of such functions has been probed in many books (see, for example, [16]). The sheaf A^0 (and all sheaves $A^{\alpha,\beta}$) possesses the following property. Let $\{f_i\}$ be a 0-cochain with values in A^0. Then the 0-cochain $\{\alpha_i f_i\}$ is also a cochain with values in A^0. Sheaves possessing this property are called *fine*. Note, for example, that the sheaf C^1 is not fine. In fact, if $\{\omega_i\}$ is the 0-cochain with values in C^1, then the forms $\omega_i \alpha_i$ cannot be closed.

THEOREM 9.1. *For any fine sheaf P, the relation*

$$H^q(M,P) = 0, \qquad q \geqslant 1 \tag{9.9}$$

is valid.

First consider the case $q = 1$. Let $\{f_{ij}\}$ be an arbitrary 1-cocycle with values in P. Examine the 0-cochain

$$f = \{f_i\} = \sum_{u_i \cap u_k \neq 0} \alpha_k f_{ik},$$

where $\sum_j \alpha_j \equiv 1$ is a partition of unity. Since $\{f_{ij}\}$ is a cocycle, its boundary is

$$\delta\{f_{ij}\} = \{f_{ij} - f_{ik} + f_{jk}\} = 0.$$

Then a coboundary of f is

$$\delta f = \{f_i - f_j\} = \left\{\sum_{k \in I} \alpha_k f_{ik} - \sum_{k \in I} \alpha_k f_{jk}\right\}$$
$$= \left\{\sum_{k \in I} \alpha_k (f_{ik} - f_{jk})\right\} = \left\{\sum_{k \in I} \alpha_k f_{ij}\right\} = \{f_{ij}\}.$$

Therefore, $\{f_{ij}\} = \delta f$. Since f is a cochain with values in P, the cocycle $\{f_{ij}\}$ is cohomological to zero, and hence $Z^1(P,M) = B^1(P,M)$, and $H^1(P,M) = 0$.

For $q > 0$ the same proof is also valid.

Taking into account (8.9), we obtain from (9.8) the exact sequences

$$\begin{gathered} 0 \to H^0(C,M) \xrightarrow{i^0} H^0(A^0,M) \xrightarrow{d^0} H^0(C^1,M) \xrightarrow{\delta_0^*} H^1(C,M) \to 0, \\ 0 \to H^q(C^1,M) \xrightarrow{\delta_q^*} H^{q+1}(C,M) \to 0. \end{gathered} \tag{9.10}$$

Therefore, δ_q^* are epimorphisms for all $q > 0$. The following theorem is valid.

THEOREM 9.2 (DE RHAM).

$$(9.11)\quad H^1(C,M) \cong H^0(C^1,M)/d^0H^0(A^0,M) = \Gamma(C^1)/d\Gamma(A^0),\quad H^q(C^1,M) \cong H^{q+1}(C,M),\quad q \geqslant 1.$$

The group $\Gamma(C^1)/d\Gamma(A^0)$ is known as the de Rham cohomology group. It is a quotient-group of closed differential forms by exact ones. Therefore, elements of this group are classes of closed forms having the same periods, and hence this group is isomorphic to C^{2g}.

Let $h \in H^1(C)$. Construct a differential form representing the class corresponding to h in the de Rham group. A 1-cocycle $\{h_{ij}\}$ corresponds to the cohomology class h, where h_{ij} are constants defined on the intersections $U_i \cap U_j$. Because $\{h_{ij}\}$ is a cocycle, its coboundary is equal to zero,

$$(9.12)\qquad h_{ij} - h_{ik} + h_{jk} = 0.$$

Let $\sum \alpha_k \equiv 1$ be a partition of unity subordinated to the covering $\{U_i\}$. Assume

$$(9.13)\qquad H_i = \sum_{u_i \cap u_k \neq 0} \alpha_k(p) h_{ik},$$

where $\{H_i\}$ is a 0-cochain with the values in A^0. From (9.12) we obtain

$$(9.14)\qquad dH_i - dH_j = d\left(\sum_k \alpha_k\right) h_{ij} = 0,$$

since h_{ij} are constants. Therefore, dH_i is a closed differential form corresponding to the cocycle h. Actually, let some differential form ω on M represent a cohomology class of the group $H^0(C^1)$. According to the construction given in the proof of Theorem 9.2, the coboundary homomorphism is built as follows. In the domain U_i, we have $\omega = d\Omega_i$, where Ω_i is a function defined in U_i. The coboundary of $\{\Omega_i\}$ is the cocycle $\{\Omega_{ij}\}, \Omega_{ij} = \Omega_i - \Omega_j$, with values in C. It is clear that in this construction the form dH_i determines the cocycle $\{h_{ij}\}$.

Introduce the operator

$$(9.15)\qquad \bar\partial : A^{\alpha,\beta} \to A^{\alpha,\beta+1}, \qquad \bar\partial = \frac{\partial}{\partial \overline{z(p)}} \wedge d\overline{z(p)}.$$

It is clear that $\bar\partial^2 = 0$ and hence the operator $\bar\partial$, can be used to construct cohomologies analogous to the de Rham ones.

First we must show that any form of type $(\alpha,1)(\alpha = 0,1)$ is locally $\bar\partial$-exact. In fact, on the unit disk $|z| < 1$, any function $f(z)$ of the class C^∞ is represented as $f(z) = \partial\varphi/\partial\bar z$, where

$$(9.16)\qquad \varphi(z) = -\frac{1}{\pi}\iint_{|t|<1} \frac{f(t)\,d\xi\,d\eta}{t-z}, \qquad t = \xi + i\eta.$$

Green's formula for the domain D_δ, obtained from $|z| < 1$ by the removal of a δ-neighbourhood of the point $t = z$, is

$$-\frac{1}{\pi}\iint\limits_{D_\delta} \frac{\partial\varphi}{\partial\bar{t}}\frac{d\xi d\eta}{t-z} = \frac{1}{2\pi i}\int\limits_{|t|=1} \frac{\varphi(t)dt}{t-z} - \frac{1}{2\pi i}\int\limits_{|t-z|=\delta} \frac{\varphi(t)dt}{t-z}.$$

Passing to the limit, we obtain

$$\varphi(z) = -\frac{1}{\pi}\iint\limits_{|t|\leqslant 1} \frac{\partial\varphi}{\partial\bar{t}}\frac{d\xi d\eta}{t-z} \quad + \text{holomorphic function},$$

as desired.

Denote by $C_z^{\alpha,0}$ the sheaf of germs of forms of the type $(\alpha, 0)$ which is closed with respect to $\bar\partial$. We obtain the exact sequence

$$0 \longrightarrow C_z^{\alpha,0} \xrightarrow{i} A^{\alpha,0} \xrightarrow{\bar\partial} A^{\alpha+1,0} \longrightarrow 0, \tag{9.17}$$

involving the exact cohomology sequence

$$H^0(A^\alpha, 0) \xrightarrow{\bar\partial^*} H^0(A^{\alpha,1}) \xrightarrow{\delta^*} H^1(C_z^{\alpha,0}) \to H^1(A^{\alpha,0}). \tag{9.18}$$

The sheaf $A^{\alpha,0}$ is fine. Hence, the last term of (9.18) is zero. We obtain the following theorem.

THEOREM 8.3 (DOLBEAULT).

$$H^1(C_z^{\alpha,0}) = H^0(A^{\alpha,1})/\bar\partial H^0(A^{\alpha,0}) = \Gamma(A^{\alpha,1})/\bar\partial\Gamma(A^{\alpha,0}). \tag{9.19}$$

The quotient-group on the right-hand side is an analog of the de Rham group for the operator $\bar\partial$. Continuing the sequence (9.18) and using the fineness of the sheaf $A^{\alpha,0}$, we obtain

$$0 \to H^q(A^{\alpha,1}) \to H^{q+1}(C_z^{\alpha,0}) \to 0. \tag{9.20}$$

This means that these groups are isomorphic. In particular, for $q > 1$, since the sheaves $A^{\alpha,1}$ are fine, it follows that

$$H^q(C_z^{\alpha,0}) = 0, \qquad \alpha = 0, 1, \qquad q > 1. \tag{9.21}$$

§10 Divisors. Complex Vector Bundles. Serre and Riemann Theorems

A. *Divisors and Complex Line Bundles*

Now we consider another definition of divisors. This definition is equivalent to the definition used in Chapter 1 but is more convenient for our purpose.

Let M^* be the multiplicative sheaf of meromorphic functions on M, and Q^* be its subsheaf of holomorphic functions different from zero. Consider the exact sequence

$$(10.1) \qquad 0 \to Q^* \xrightarrow{i} M^* \xrightarrow{j} \theta \to 0,$$

where $\theta = M^*/Q^*$ is a quotient-sheaf with the addition as a group operation. The sheaf θ is called a *sheaf of germs of divisors*, and its sections over M are called *divisors*. The additive group $H^0(\theta) = \Gamma(\theta, M)$ is called the *divisor group of the surface M*.

In order to compare both definitions, consider some section d of the sheaf θ over a simply-connected domain U. By definition of a quotient-sheaf, such a section determines some element of the group $\Gamma(M^*, U)/\Gamma(Q^*, U)$. Let this element be determined by the element of the group $\Gamma(M^*, U)$, given by a meromorphic function $m_U(p)$ in U with an order $\alpha(p)$ at the point p (recall that $\alpha(p)$ is equal to the order of zero or to the order of a pole with an opposite sign). We obtain the relation $d : p \to \alpha(p), \alpha(p)$ being zero everywhere except for a finite set of points. Conversely, every such relation, of course, defines the class of elements of the groups $\Gamma(M^*, U)$ corresponding to an element of the group $H^0(\theta)$.

Divisors are closely associated with holomorphic bundles with fibre C (complex line bundles). The holomorphic vector bundle B over M, with a structure group $GL(n, C)$ and fibre C^n, is a topological space for which the following objects are defined:

a) A mapping $\pi : B \to M$ continue in topology of B called the projection.

b) For every point $b \in B$, $\pi(b) = p$, and for every domain $U_i \ni p$ belonging to some fixed open covering $\{U_i\}$ of the surface M, there exists some vector $h_i \in C^n$, such that the pair (p, h_i) would determine the point b.

c) The system of holomorphic $n \times n$-matrices $h_{ij}(p)$, called the *transition matrices*, defined in the intersections $U_i \cap U_j$, is nondegenerate, and satisfies

$$(10.2) \qquad h_{ij} h_{jk} h_{ki} = E,$$

where E is the unit matrix, and at every point p,

$$(10.3) \qquad h_i = h_{ij}(p) h_j.$$

The bundles B and B_1 with transition matrices $\{h_{ij}\}$ and $\{h'_{ij}\}$ are called equivalent if there exists such a system of nondegenerate $n \times n$-matrices $h_i(p)$ defined in U_i, so that

$$(10.4) \qquad h'_{ij}(p) = h_i(p) h_{ij}(p) h_j^{-1}(p).$$

A direct sum of bundles B and B' with fibres C^n and C^{n_1} and transition matrices $\{h_{ij}\}$ and $\{h'_{ij}\}$ is the bundle $B \bigoplus B'$ with fibre C^{n+n_1} the group $GL(n+n_1, C)$ and transition matrices

$$\tilde{h}_{ij} = \begin{pmatrix} h_{ij} & 0 \\ 0 & h'_{ij} \end{pmatrix}. \tag{10.5}$$

In the case $n = 1$, we have a bundle whose fibres are complex lines (such bundles are called *complex line bundles*). The sum $B + B'$ of the line bundles is the line bundle with the transition functions[2]

$$\tilde{h}_{ij} = h_{ij} h'_{ij}. \tag{10.6}$$

Consider the group L of classes of equivalent line bundles with the operation (10.6). Every line bundle defines a cocycle $\{h_{ij}\}$ with values in Q^*. Equivalent bundles correspond, on account of (10.4), to cohomology cocycles of transition functions. Therefore, for every element of the group L, there exists a cohomology class belonging to the group $H'(Q^*)$. This involves the isomorphism $L = H^1(Q^*)$.

The sequence (10.1) generates the cohomology sequence

$$H^0(M^*) \xrightarrow{j^*} H^0(\theta) \xrightarrow{\delta^*} H^1(Q^*) \to L. \tag{10.7}$$

Therefore, every divisor on M determines a class of equivalent bundles. In order to realize this correspondence, it is necessary to construct the homomorphism δ^*. This may be made in the following manner. Let the divisor h correspond to the 0-cochain $\{h_i(p)\} \in Cn^0(M^*)$. The homomorphism δ^* transforms the cohomology class represented by h in the cohomology class of $H^1(Q)$ determined by the cocycle $\{h_{ij}\} \in Z^1(Q^*)$, to

$$h_{ij}(p) = \frac{h_i(p)}{h_j(p)}.$$

Taking $\{h_{ij}\}$ as transition functions, we can define the bundle B_h corresponding to the divisor h. The kernel of this mapping is formed by divisors h, such that the cocycles $\{h_{ij}\}$ are cohomological to zero. This means that there exists such a 0-cochain $\{h'_i\} \in C^0n(Q^*)$ that $h_{ij} = h'_i(h'_j)^{-1}$. Therefore, taking into account the definition of $\{h_{ij}\}$, we obtain $h_i(p)(h'_i(p))^{-1} = h_j(p)(h'_j(p))^{-1} = h(p)$, which is a function defined on all surfaces M. The divisor of this function is h. Therefore, in order that $\delta^*(h) = 0$, it is necessary and sufficient that the divisor h define a function meromorphic on M. On the other hand, the triviality of the cocycle $\{h_{ij}\}$ is equivalent to the relation $h'_i = h_{ij} h'_j$. Hence $\{h'_i\}$ is, in this case, a holomorphic section of the bundle B_h. Therefore, for the bundle B to be trivial it is necessary and sufficient that the group of sections $\Gamma(M, B)$ be nonempty.

---

[2]For n-dimensional bundles, this notion corresponds to a tensor product of bundles [5].

The bundles B_h are the most important example of complex line bundles. Below we shall show that (10.7) defines the isomorphism of the group of classes of equivalent divisors and the group L (and hence the bundles B_h exhaust all complex line bundles).

In some cases, line bundles are determined by transition functions. In particular, this situation occurs in the study of values depending on a choice of a local coordinate. It is convenient to consider them as sections of some bundles.

Consider the so-called canonical bundle K whose transition functions are determined by $f_{ij} = dz_j/dz_i$ in the domains $U_i \cap U_j$. Sections $\{g_i\}$ of this bundle are called covariants. It is clear that $g_i dz_i$ are differential forms of type (1,0).

Consider some sheaves associated with line bundles. The first of them is the sheaf of germs of holomorphic sections of a bundle B. We denote it by $Q(B)$. In particular, the sheaf $Q(B_h)$ coincides with the sheaf Q_h of germs of meromorphic functions which are multiples of the divisor $-h$. In fact, if $f_i(p)$ is a section of B_h over the domain U_i, then $f_i = h_{ij} f_j$ in $U_i \cap U_j$. Then $\{f_i h_i^{-1}\}$ is a section of Q_h, and conversely. The sheaf $Q(K)$ coincides with the sheaf Q^1 of germs of holomorphic forms of the type (1,0).

More general examples of sheaves are associated with a notion of a differential form with coefficients in a bundle. The coefficients of a differential form $\omega = a_i dz_i + b_i d\bar{z}_i \in H^0(A^1)$ can be considered as sections of the bundles K and $\overline{K}$ with the transition functions $f_{ij} = dz_j/dz_i$ and $\bar{f}_{ij} = d\bar{z}_j/d\bar{z}_i$.

Let now a_i and b_i be sections of the bundles $B + K$ and $B + \overline{K}$, respectively. Then, by definition, the form $\omega = a_i dz_i(p) + b_i \overline{dz_i(p)}$ is a 1-form with values in B.

If ω_1 and ω_2 are 1-forms with values in B_1 and B_2, respectively, then by definition, $\omega_1 \wedge \omega_2$ is a 2-form with values in $B_1 + B_2$.

We shall examine the sheaves $A^{\alpha,\beta}(B)$, $C^{\alpha,\beta}(B)$, and so on, of forms of type (α, β) with values in B.

It is easy to verify that the sheaves $A^{\alpha,\beta}(B)$ are fine, and that the sequence

$$(10.8) \qquad 0 \to C_z^{\alpha,\beta}(B) \to A^{\alpha,\beta}(B) \xrightarrow{\bar{\partial}} A^{\alpha,\beta+1}(B) \to 0$$

is exact. From (10.8) we obtain the Dolbeault theorem of the form

$$(10.9) \qquad H^1(C_z^{\alpha,0}) = \Gamma(A^{\alpha,1}(B))/\bar{\partial}\Gamma(A^{\alpha,0}(B))$$

$$(10.10) \qquad H^q(C_z^{\alpha,0}(B)) = 0, \qquad q > 1.$$

B. *The Serre Duality Theorem*

THEOREM 10.2 (SERRE). *The groups $H^q(Q^\alpha(B))$ and $H^{1-q}(Q^{1-\alpha}(-B))$, $q, \alpha = 0, 1$ are dual. These groups are finite dimensional as vector spaces, and hence*

$$(10.11) \qquad \dim H^q(Q^\alpha(B)) = \dim H^{1-q}(Q^{1-\alpha}(-B)), \qquad q, \alpha = 0, 1.$$

By symmetry, it is sufficient to examine the case $q = 1$. Let $\tilde{A}^{\alpha,\beta}(B)$ be a sheaf of germs of forms of type (α, β) with values in the line bundle. We suppose that coefficients of these forms are distributions (generalized functions; see [21]). Consider the diagram

$$
(10.12)\qquad
\begin{array}{ccc}
\Gamma(A^{\alpha,0}(B)) & \xrightarrow{\bar\partial} & \Gamma(A^{\alpha,1}(B)) \\
\downarrow & & \downarrow \\
\Gamma(\widetilde{A}^{1-\alpha,1}(-B)) & \xleftarrow{\bar\partial^*} & \Gamma(\widetilde{A}^{1-\alpha,0}(-B)).
\end{array}
$$

Here the vertical arrows denote the transitions to the dual spaces. Corresponding bilinear forms on pair of spaces $\Gamma(A^{\alpha,\beta}(B))$ and $(\widetilde{A}^{1-\alpha,1-\beta}(-B))$ are determined by the integrals

$$
(10.13)\qquad <\omega,\eta>=\int_M \omega\wedge\eta,\quad \omega\in\Gamma(A^{\alpha,\beta}(B)),\quad \eta\in\Gamma(\tilde{A}^{1-\alpha,1-\beta}(-B)).
$$

It is clear that the integral (10.13) is defined properly.

The operator $\bar\partial^*$ is dual to $\bar\partial$. As is well known, these Cauchy–Riemann operators are elliptic (see [21]). The ellipticity of $\bar\partial$ means that the zero space of the dual operator

$$\operatorname{Ker}\bar\partial^* = \Gamma(\widetilde{C}_z^{1-\alpha,0}(-B))$$

is finite dimensional. The kernel of $\bar\partial^*$ is formed by the holomorphic forms

$$\widetilde{C}_z^{1-\alpha,0}(-B) = C_z^{1-\alpha,0}(-B) = Q^{1-\alpha}(-B).$$

Therefore,

$$
(10.14)\qquad \operatorname{Ker}\bar\partial^* = H^0(Q^{1-\alpha}(-B)).
$$

The space (10.14) is dual to the cokernel of the operator $\bar\partial$,

$$
(10.15)\qquad \operatorname{Co\,ker}\bar\partial = \Gamma(A^{\alpha,1}(B)/\bar\partial\Gamma(A^{\alpha,0}(B)).
$$

With respect to the Dolbeault theorem,

$$
(10.16)\qquad \operatorname{Co\,ker}\bar\partial \cong H^1(Q^\alpha(B)),
$$

which completes the proof.

Note. The simplicity of our proof is provided by using well-known but very non-trivial properties of elliptic operators. A direct formal proof which is independent of the theory of elliptic operators is available in [31].

C. *The Riemann Theorem*

Here we give a new proof of the Riemann theorem (Theorem 2.4) concerning the number of Abelian differentials of the first kind.

THEOREM 10.3.

$$\dim H^0(Q^1) = g. \tag{10.17}$$

PROOF: If F is a sheaf, the Euler characteristic of this sheaf is defined as the alternating sum of dimensions of cohomology groups

$$\chi(F) = \sum_{q=0}^{\infty}(-1)^q \dim H^q(F, M). \tag{10.18}$$

Passing from the exact sequence

$$0 \to F' \to F \to F'' \to 0$$

to the cohomology sequence

$$\cdots \to H^q(F') \to H^q(F) \to H^q(F'') \to \dots,$$

we obtain the relation well-known in homological algebra [5],

$$\chi(F) = \chi(F') + \chi(F''). \tag{10.19}$$

Therefore, the exact sequence

$$0 \to C \to Q \xrightarrow{d} Q^1 \to 0 \tag{10.20}$$

involves the relation

$$\chi(Q) = \chi(C) + \chi(Q^1). \tag{10.21}$$

Calculate the value of the right-hand side of (10.21). First we examine the constant sheaf C of complex numbers. We have $\dim H^0(C) = \dim \Gamma(C) = 1$ (the dimension is complex). Further, on account of (9.11),

$$\dim H^1(C) = \dim H^0(C^1)/d\Gamma(A^0) = 2g. \tag{10.22}$$

Finally, $H^{q+1}(C) = H^q(C^1), q \geqslant 1$. In order to calculate the dimensions of these groups, consider the exact sequence

$$0 \to C^1 \to A^1 \xrightarrow{d} A^2 \to 0. \tag{10.23}$$

With regard to the fineness of the sheaves A^1 and A^2, (10.23) involves the exact sequence

$$\begin{gathered} 0 \to H^0(C^1) \to H^0(A^1) \xrightarrow{d} H^0(A^2) \to H^1(C^1) \to 0, \\ 0 \to H^q(A^2) \to H^{q+1}(C^1) \to 0, \qquad q \geqslant 1. \end{gathered} \tag{10.24}$$

This means that

$$(10.25)\qquad \begin{aligned} H^1(C^1) &\cong H^0(A^2)/dH^0(A^1),\\ H^{q+1}(C^1) &\cong H^q(A^2) = 0, \qquad q \geqslant 1. \end{aligned}$$

Every element of the group $H^0(A^2)/dH^0(A^1)$ defines the number

$$(10.26)\qquad c = \int_M \eta,$$

where $\eta \in H^0(A^2)$ is a form representing the corresponding cocycle, since for any 1-form ω,

$$\int_M d\omega = \int_{\partial M} \omega = 0.$$

Therefore, we have the isomorphism

$$H^1(C^1) \cong H^0(A^2)/dH^0(A^1) \cong C$$

and

$$\dim H^2(C) = 1$$

(see (9.11). From (10.25) we obtain that

$$\dim H^{q+1}(C) = \dim H^q(C^1) = 0, \qquad q > 1.$$

Thus, we have the relation

$$(10.27)\qquad \chi(C) = 2 - 2g.$$

Further, from the Serre theorem (10.11), it follows that

$$\dim H^1(Q^1) = \dim H^0(Q^0) = 1.$$

From (9.21), $H^2(Q^1) = 0$, and we obtain

$$(10.28)\qquad \chi(Q^1) = -1 + \dim H^0(Q^1).$$

We proceed with a calculation of the value $\chi(Q)$. The Liouville theorem involves $H^0(Q) = C$. Further, by the Serre theorem and (9.21), $\dim H^2(Q) = \dim H^2(Q^1) = 0, \dim H^1(Q) = \dim H^0(Q^1)$. We have

$$(10.29)\qquad \chi(Q) = 1 - \dim H^0(Q^1).$$

From (10.21), (10,27), (10.28), and (10.29) we obtain

$$(10.30)\qquad 1 - \dim H^0(Q^1) = 2 - 2g + \dim H^0(Q^1) - 1.$$

The relation (10.30) involves (10.17) and an important corollary,

$$\chi(Q) = 1 - g. \tag{10.31}$$

§11 The Riemann–Roch Theorem. The Riemann Problem

A. *The Riemann–Roch Theorem*

Let $h \geqslant 0$ be a divisor. Consider the exact sequence

$$0 \to Q_h \xrightarrow{i} Q \to Q(h) \to 0. \tag{11.1}$$

Here Q_h is a sheaf of germs of holomorphic functions which are multiples of the divisor h; $Q(h) = Q/Q_h$ is a quotient-sheaf determined in the domains U_i by the groups Q_i/Q_{ih} of holomorphic group functions in U_i by the subgroups Q_{ih} of functions which are multiples of the divisors $h \cap U_i$. If the covering $\{U_i\}$ is sufficiently refined, every domain U_i contains no more than one point of the divisor h. Thus, the group Q/Q_{ih} is trivial or isomorphic to C^{α_i}, where α_i is an order of zero determined by the divisor h at the point $p_i \in U_i$ (an element of C^{α_i} is determined by the first α_i coefficients of the Taylor expansion of the function belonging to Q_i at p_i).

It is clear that

$$\dim H^0(Q(h)) = \deg h. \tag{11.2}$$

On the other hand, the covering may be refined such that every intersection $U_i \cap U_j$ contains no point of the divisor h, and hence

$$\dim H^q(Q(h)) = 0, \qquad q \geqslant 1. \tag{11.2'}$$

Because of (10.9), we have

$$\chi(Q_h) + \chi(Q(h)) = \chi(Q). \tag{11.3}$$

From (10.31), $\chi(Q) = 1 - g$. Taking into account (11.2), (11.2′), and (11.3), we obtain the relation

$$\chi(Q_h) = 1 - \deg h - g. \tag{11.4}$$

An arbitrary divisor h can be represented by

$$h = h_0 - h_1, \qquad h_0, h_1 \geqslant 0.$$

The sequence

$$0 \to Q_{h_0} \to Q_h \to Q(h_1) \to 0 \tag{11.5}$$

is exact. From (11.4) it follows that

$$\begin{aligned}(11.6)\quad \chi(Q_h) &= \chi(Q_{h_0}) + \chi(Q(h_1)) \\ &= \chi(Q_{h_0}) + \deg h_1 = 1 - g - \deg h_0 + \deg h_1.\end{aligned}$$

Calculate the left side of Eq. (11.6). On account of (10.10),

$$\dim H^q(Q_h) = \dim H^q(Q(-B_h)) = 0, \qquad q > 1.$$

Thus,

$$(11.7)\qquad \dim H^0(Q_h) - \dim H^1(Q_h) = 1 - \deg h - g.$$

Considering that $H^0(Q_h)$ is the group of analytic functions which are multiples of the divisor h, and that according to the Serre theorem,

$$H^1(Q_h) = H^1(Q^0(B_{-h})) = H^0(Q^1(B_h)) = H^0(Q^1_{-h}),$$

we obtain the Riemann–Roch theorem

$$(11.8)\qquad \dim H^0(Q_h) - \dim H^0(Q^1_{-h}) = 1 - g - \deg h.$$

Note. If we write $-h$ in place of h, we obtain

$$(11.8')\qquad \dim H^0(Q_{-h}) - \dim H^0(Q^1_h) = \deg h - g + 1.$$

B. *Some Corollaries*

Let $q_1 q_2 \in M$ be points on the surface and the divisor $h = -(q_1 + q_2)$. Then

$$(11.9)\qquad \dim H^0(Q^1_h) = g + 1 + \dim H^0(Q_{-h}).$$

As was shown in §10, the number of Abelian differentials of the first kind is equal to g. From (11.9) it follows that $\dim H^0(Q^1_h) > g$ and hence that there exists an Abelian differential of the third kind with poles of the first order at the points q_1 and q_2. Assuming $h = nq, n > 1$, we can prove in the same manner the existence of the Abelian differential of the second kind with a pole at q.

Therefore, we proved again the existence of Abelian differentials of the first, second, and third kind.

Note. In our case the Weyl lemma used in Chapter 1 and the ellipticity of the Cauchy–Riemann operator $\bar\partial$ used in Chapter 2 are equivalent.

THEOREM 11.1. *The homomorphism (10.7) generates an isomorphism between the groups $H^0(\theta)/j^* H^0(M^*)$ and $H^1(Q^*)$.*

Let us show that every line bundle B with transition functions $\{h_{ij}\}$ corresponds to some divisor.

Let $h_0 \geqslant 0$ be some divisor. The exact sequence

$$(11.10)\qquad 0 \to Q(B - B_{h_0}) \to Q(B) \to h_0 \times C \to 0$$

generates the exact sequence

$$H^0(Q(B)) \to H^0(h_0 \times C) \to H^1(Q(B - B_{h_0})). \tag{11.11}$$

The middle term of (11.11) is different from zero and hence one of the other terms is also nonzero.

Let $H^0(Q(B)) \neq 0$ and $\{h_i\}$ be a section of $Q(B)$. Then $h_i = h_{ij}h_j$. This means that the bundle B determines the divisor $\{h_i\}$. If the sheaf $Q(B)$ has no holomorphic sections, then the group $H^1(Q(B-B_{h_0}))$ is different from zero. Due to the Serre theorem, this group is isomorphic to the group $H^0(Q^1(-B+B_{h_0}))$. Therefore the sheaf $Q^1(-B+B_{h_0})$ has a holomorphic section $\{f_i\}$,

$$f_i = h_{ij}^{-1} \frac{(h_0)_i}{(h_0)_j} \frac{dz_j}{dz_i} f_j.$$

Let $\eta = \eta_i dz_i$ be some Abelian differential. Then

$$h_{ij} = \frac{(h_0)_i \eta_i f_i^{-1}}{(h_0)_j \eta_j f_j^{-1}},$$

and consequently the bundle B defines the divisor

$$h = h_0 + \{\eta_i f_i^{-1}\}.$$

In conclusion, examine the operator $\bar\partial$ on a sheaf A^0_{-h} of germs of functions which are multiples of the divisor $-h$. In general, as long as a differentiation increases pole orders, we shall consider the sheaf $A^0(B_h)$ of germs of sections of the bundle B_h instead of A^0_{-h}. Consider the diagram

$$\begin{array}{ccc} A^0(B_h) & \xrightarrow{\ \bar\partial\ } & A^{0,1}(B_h) \\ \downarrow & & \downarrow \\ \tilde{A}^{1,1}(-B_h) & \xleftarrow{\ \bar\partial^*\ } & \tilde{A}^{1,0}(-B_h) \end{array} \tag{11.12}$$

(cf. (10.12)), where $\tilde{A}^{\alpha,\beta}(B)$ are sheaves of germs of differential forms with values in B whose coefficients are distributions. The vertical arrows denote the transitions to the dual spaces, and duality is determined by the bilinear form

$$< \omega, \eta >= \int_M \omega \wedge \eta. \tag{11.13}$$

Note that the values described by the Riemann–Roch theorem are closely related to the diagram (11.12). In particular, $H^0(Q(B_h)) = \operatorname{Ker} \bar\partial$, and $H^0(Q^1(B_{-h})) = \operatorname{Ker} \bar\partial^*$. Therefore, the Riemann–Roch theorem is an index theorem for the Cauchy–Riemann operator $\bar\partial$.

C. *The Riemann Boundary Problem*

In this section we establish the relation between the Riemann boundary problem and complex vector bundles. This relation was shown by Grothendieck [a] for the complex sphere and by Röhrl [c] for closed Riemann surfaces.

Consider the Riemann problem

$$F^+(p) = G(p)F^-(p) \tag{11.14}$$

and conjugate problem for differentials (see §5).

$$f^+(p)dz(p) = \frac{1}{G(p)} f^-(p)dz(p). \tag{11.15}$$

Let $\{\varphi_i\}$ be a 0-cochain where φ_i are solutions (11.14) in the domains U_i different from zero. Therefore, $\varphi_i(p)$ are defined in the domains U_i, different from zero at all points and satisfying the condition (11.14) on $L \cap U_i$ (if $L \cap U_i$ is empty, φ_i is holomorphic in U_i). Define the complex line bundle B_G by the transition functions

$$g_{ij} = \frac{\varphi_j(p)}{\varphi_i(p)}, \qquad p \in U_i \cap U_j. \tag{11.16}$$

THEOREM 11.2. *For every complex line bundle B over the compact Riemann surface M, there exists an equivalent bundle B_G of the form (11.16).*

We refer to Theorem 14.9 which will be proved below: In any domain of a Riemann surface bounded by a smooth curve, every 1-cocycle $\{h_{ij}\}$ with values in Q^* is trivial and may be represented by $h_{ij} = h_i h_j^{-1}$, where $\{h_i\}$ is a cochain with values in Q^*. We have such a representation in both domains $T^\pm$. If the domain U_i is separated into two parts U_i^+ and U_i^- by the contour L, we have the representation

$$h_{ij}(p) = \begin{cases} h_i^+(p)(h_j^+(p))^{-1} & p \in T^+ \cap U_i \cap U_j \\ h_i^-(p)(h_j^-(p))^{-1} & p \in T^- \cap U_i \cap U_j. \end{cases}$$

On the contour L, we have

$$\frac{h_i^+(p)}{h_i^-(p)} = \frac{h_j^+(p)}{h_j^-(p)} = G(p), \qquad p \in L.$$

We assume

$$\varphi_i(p) = \begin{cases} h_i^+(p) & p \in U_i \cap T^+ \\ h_i^-(p) & p \in U_i \cap T^-. \end{cases} \tag{11.17}$$

Therefore, the classification of line bundles may be made in two ways: by divisors (Theorem 11.1) and by Riemann problems. Of course, both methods are equivalent. This fact causes an interest in the Riemann problem. This approach provides new possibilities for the study of the Riemann problem.

THEOREM 11.3. *The boundary problem (11.14) has a solution with a finite number of poles.*

Consider the bundle B_G. Due to Theorem 11.1, there exists a divisor $\bar{h}$, such that B_G is equivalent to $B_{\bar{h}}$. This means that

$$g_{ij} = h'_i h_{ij} h'^{-1}_j \quad \text{in} \quad U_i \cap U_j, \tag{11.18}$$

where h'_i are holomorphic and different from zero. If the divisor $\bar{h}$ is determined by the meromorphic cochain $\{h_i\}$, then it follows from (11.18) that

$$\varphi_i(p)h_i(p)h'_i(p) = \varphi_j(p)h_j(p)h'_j(p) = F^{\pm}(p), \qquad p \in U_i \cap U_j \tag{11.19}$$

is a solution of (11.14), which is a multiple of the divisor $\bar{h}$.

Every holomorphic solution of the problem (11.14) corresponds to a holomorphic section of the bundle B_G. Hence the number of the solutions of problem(11.14) is equal to $l = \dim H^0(Q(B_G)) = \dim H^0(Q(B_{\bar{h}}))$.

Every holomorphic solution of problem (11.15) corresponds to a holomorphic section of the sheaf $Q(-B_G + K)$ (for a definition of K, see p.62) and conversely. In fact, if $g = \{g_i\}$ is a section of the sheaf $Q(-B_G + K)$, then $g_i = g_{ij}g_j$, where

$$g_{ij} = \frac{\varphi_i(p)}{\varphi_j(p)} \frac{dz_j}{dz_i}.$$

This means that

$$\frac{g_i dz_i}{\phi_i} = \frac{g_j dz_j}{\phi_j} = f^{\pm}(\rho)dz(\rho)$$

is a solution of problem(11.15). The number of solutions of problem (11.15) is therefore equal to

$$h = \dim H^0(Q(-B_G + K)) = \dim H^0(Q^1(-B_G)) = \dim H^1(Q^0(-B_{\bar{h}})).$$

Using the Riemann–Roch theorem, we obtain

$$l - h = \deg \bar{h} - g + 1. \tag{11.20}$$

Because any line bundle $B_{\bar{h}}$ corresponds to a bundle B_G, this relation is equivalent to the Riemann–Roch theorem.

Solvability of the Riemann problem will be studied in the following sections.

Note that $\deg h$ is equal to the index of the boundary condition. In fact, if the divisor $\bar{h}$ is such that $B_{\bar{h}}$ is equivalent to B_G, then φ_i/h_i is a solution of the problem (11.14), and $\deg(\varphi_i/h_i) = \kappa = \operatorname{ind}_L G$.

§12 The Second Cousin Problem. Solvability of the Riemann Problem

A. *Characteristic Classes. Abel's Theorem*

Consider the exact sequence

$$O \to J \xrightarrow{i} Q \xrightarrow{e} Q^* \to 0. \tag{12.1}$$

Here J is the constant sheaf of integers, and $e(f) = \exp 2\pi i f(p)$. It follows from the Liouville theorem, that the homomorphism $H^0(Q) \xrightarrow{e^*} H^0(Q^*)$ is an epimorphism. Hence we have the exact sequence

$$0 \to H^1(J) \xrightarrow{i^*} H^1(Q) \xrightarrow{e^*} H^1(Q^*) \xrightarrow{\nu} H^2(J). \tag{12.2}$$

If L denotes the group of classes of equivalent complex line bundles, then the isomorphism $g : L \to H^1(Q^*)$ induces the homomorphism $c = \nu g : L \to H^2(J)$ which is called *characteristic*. Therefore, every complex line bundle B determines the integer cohomology class $c(B)$ called a *characteristic class* or a *Chern class*. Let $\{h_{ij}\}$ be the transition functions determining the cohomology class $h \in H^1(Q^*)$. Calculate the class νh. It is determined by the cocycle $\{g_{ijk}\}$,

$$g_{ijk} = \frac{1}{2\pi i}(\ln h_{ij} + \ln h_{jk} - \ln h_{ik}), \qquad p \in U_i \cap U_j \cap U_k. \tag{12.3}$$

According to the de Rham theorem, this class determines a 2-form, which we have to calculate.

The de Rham sequence

$$0 \to C \xrightarrow{i} A^0 \xrightarrow{d} C^1 \to 0 \tag{12.4}$$

generates the isomorphism

$$0 \to H^1(C^1) \xrightarrow{\delta^1} H^2(C) \to 0. \tag{12.5}$$

Denote by $\hat{h}$ the cohomology class of the group $H^2(C)$ whose preimage is determined by the cocycle $\{g_{ijk}\} \in Z^2(C)$. It is clear that the class $\hat{h} \in H^2(C)$ is determined by the cocycles $\{\frac{1}{2\pi i} d \ln h_{ij}\} \in Z^1(C^1)$. Consider the exact sequence

$$0 \to C^1 \xrightarrow{i} A \xrightarrow{d} A^1 \to 0. \tag{12.6}$$

It induces the sequence

$$H^0(A^1) \xrightarrow{d^*} H^0(A^2) \xrightarrow{\delta^*} H^1(C^1). \tag{12.7}$$

Calculate $\delta^{*-1}\hat{h}$. Let $\sum \alpha_k \equiv 1$ be a partition of unity. Assume

$$T_i(p) = \frac{1}{2\pi i} \sum_{k \in I} \alpha_k d \ln h_{ik}(p). \tag{12.8}$$

From (12.3) we have

$$T_i - T_j = \frac{1}{2\pi i}\sum_{k\in I}\alpha_k d(\ln h_{ik} - \ln h_{jk}) = \frac{1}{2\pi i}\ln h_{ij}. \tag{12.9}$$

Taking into account the rule of the coboundary homomorphism construction, we obtain the class $\delta^{*-1}\hat{h}$ which is represented by

$$\delta^{*-1}\hat{h} \sim dT_i = \frac{1}{2\pi i}\sum_{k\in I}\alpha_k d\ln h_{ik}. \tag{12.10}$$

This form represents the characteristic class in the group $\Gamma(A^2)/d\Gamma(A^1)$.

Consider the diagram

$$\begin{array}{ccc} \Gamma(A^1) & \xrightarrow{\ d\ } & \Gamma(A^2) \\ \downarrow & & \downarrow \\ \Gamma(\widetilde{A}^1) & \xleftarrow{\ d^*\ } & \Gamma(\widetilde{A}^0), \end{array} \tag{12.11}$$

where the vertical arrows are the transitions to dual spaces, and $\widetilde{A}^\alpha$ are the sheaves of germs of those forms whose coefficients are distributions. The group $\operatorname{Coker} d = \Gamma(A^2)/d\Gamma(A^1)$ is dual to the group $\operatorname{Ker} d^* = C$ and hence may be realized as a group of functionals on C. We associate the functional

$$< \omega, c >= c\int_M \omega \tag{12.12}$$

with any $\omega \in \Gamma(A^2)$. This functional vanishes on the group $d\Gamma(A^1)$, i.e., for every $\eta \in \Gamma(A^1)$,

$$\int_M d\eta = \int_{\partial M} \eta = 0.$$

Conversely, if $< \omega, 1 >= 0$, then $\omega \in d\Gamma(A^1)$. In fact, in this case the differential

$$\eta(q) = \frac{1}{\pi}\int_M \omega(p) \wedge M(q,p)dz(p) \wedge dz(q), \tag{12.13}$$

where $M(q,p)$ is the Cauchy kernel (4.20), possesses the following properties

a) $\eta(q) \in \Gamma(A^1)$.

b) $\bar{\partial}\eta = \omega$.

This follows from (9.16) on $M - \delta$, where δ is the characteristic divisor of the kernel. At the points δ of the integral, (12.13) converges, and hence the relation is valid at this point, too.

c) $\partial\eta = 0$, since η has the type (1,0). This involves the relation

$$d\eta = \partial\eta + \bar{\partial}\eta = \omega. \tag{12.14}$$

d) The regularity of η at the point q_0 follows from Eq. (4.26) and the relation $< \omega, 1 > \nu = 0$.

Choose some subset $I_0 \subset I$ of indices such that $\{U_i\}, i \in I_0$ is a triangulation (we suppose that points of the divisor are interior points of triangles). Then

$$\int_M dT_i = \sum_{i \in I_0} \int_{U_i} dT_i = \sum_{i \in I_0} \int_{\partial U_i} T_i = \sum_{i \in I_0} \frac{1}{2\pi i} \int_{\partial U_i} \sum_{k \in I} \alpha_k d \ln(h_i/h_k)$$
$$= \sum_{i \in I_0} \frac{1}{2\pi i} \int_{\partial U_i} d \ln h_i - \frac{1}{2\pi i} \sum_{i \in I_0} \int_{\partial U_i} \sum_{k \in I} \alpha_k d \ln h_k. \tag{12.15}$$

The first integral in (12.15) is calculated by the residue theorem. It is clear that it is equal to $\deg h$. The second term is equal to zero, since the 1-form

$$\sum_{k \in I} a_k d \ln h_k$$

is integrated over 1-simplexes ∂U_i, and every simplex is passed twice in opposite directions. Therefore, we obtain the important result:

THEOREM 12.1. *Let $c : L \to H^2(J)$ be the characteristic homomorphism. Then the class $c(B)$ corresponds to the differential form (12.10) representing the equivalence class of the factor-group $\Gamma(A^2)/d\Gamma(A^1)$ and to the functional (12.12) on C and*

$$< dT_i, 1 >= \deg h(B). \tag{12.16}$$

Here $h(B)$ is the divisor corresponding to B, such that $B_h \sim B$.

The idea of our construction is clear: If $c(B)$ is a cohomology class, then the functional (12.12) on the homology group $H_2(M)$ is defined, and $\deg h(B)$ is the value of this functional on the single 2-cycle of the group. This cycle is M.

B. *The Second Cousin Problem*

Let h be a divisor on M determined by the cochain $\{h_i(p)\}$. Ascertain the conditions of the existence of the analytical function $f(p)$ such that $(f) = h$. This problem was studied in §6 and was called *Abel's problem.* The other traditional name of this problem from the theory of functions of several complex variables is the *second Cousin problem.*

Consider the exact sequence

$$0 \longrightarrow Q^* \xrightarrow{i} M^* \xrightarrow{j} \theta \longrightarrow 0 \tag{12.17}$$

(see (10.1)). We obtain the exact sequence

$$H^0(M^*) \xrightarrow{j^*} H^0(\theta) \xrightarrow{\delta^*} H^1(Q^*). \tag{12.18}$$

In order to solve the second Cousin problem, it is necessary and sufficient that $h \in j^* H^0(M^*)$, i.e.,

$$\delta^* h = 0. \tag{12.19}$$

On account of (12.12), the condition (12.19) involves the condition $\nu(\delta^* h) = 0$, i.e., $c(B_h) = \deg h = 0$. This necessary condition is assumed to be valid. From (12.12) it follows that in this case the cohomology class of the group $H^1(Q^*)$ determined by the cocycle $\{h_i/h_j\}$ belongs to $e^* H^1(Q)$. For $\delta^* h = 0$ it is necessary and sufficient that $e^{*-1}(\delta^* h) \in i^* H^1(J)$. The class $e^{*-1}(\delta^* h)$ is determined by the cocycle

$$\{\frac{1}{2\pi i} \ln(h_i/h_j)\} \in Z^1(Q). \tag{12.20}$$

The Dolbeault theorem determines the isomorphism $H^1(Q) \to \Gamma(A^{0,1})/\bar{\partial}\Gamma(A^{0,0})$, sending the cocycle (12.20) to the equivalence class corresponding to the form $\bar{\partial} t_i$,

$$t_i = \frac{1}{2\pi i} \sum_{k \in I} \alpha_k \ln h_{ik}. \tag{12.21}$$

Here it is necessary to take into account that in this case the coboundary (12.3) of the cocycle $\{\frac{1}{2\pi i} \ln h_{ij}\}$ is cohomological to zero. Because of the Serre duality, the class associated with the form $\bar{\partial} t_i$ determines the functional on the group $H^0(Q^1)$,

$$(\bar{\partial} t_i, \varphi) = \int_M \bar{\partial} t_i \wedge \varphi, \qquad \varphi \in H^0(Q^1). \tag{12.22}$$

The differential form $\bar{\partial} t_i$ can be represented in another form which is more convenient for the following.

Examine the integral

$$g_1(p) = -\frac{1}{2\pi i} \int_{\widehat{M}} \bar{\partial} t_i(q) \wedge M^*(q,p) dz(q). \tag{12.23}$$

Here $M^*(p,q)$ is the multivalued Cauchy kernel (4.3), and $\widehat{M}$ is the canonical cut surface M. It is clear that $g_1(p)$ is a multivalued function whose differential $dg_1 \in H^1(C^1)$, and $\bar{\partial} g_1 = \bar{\partial} t_i(p)$ (see (9.16)).

Adding to $g_1(p)$ a combination of Abelian integrals of the first kind $dW_j (j = 1, \ldots, 2g)$, we obtain the function $g(p)$ possessing the following properties,

$$\begin{aligned} &\text{a) } dg \in H^0(C^1), \text{all periods of } g \text{ are real,} \\ &\text{b) } \bar{\partial} g = \bar{\partial} t_i. \end{aligned} \tag{12.24}$$

It is clear that these conditions determine the integral $g(p)$ completely.

We obtain (cf. Eq. (3.5)) the relation

$$(12.25)\quad \int_{\widehat{M}} \bar{\partial} t_i \wedge \varphi = \int_{\widehat{M}} \bar{\partial} g \wedge \varphi = \int_{\partial \widehat{M}} g\varphi$$

$$= \sum_{\mu=1}^{g} (c_{2\mu} \int_{k_{2\mu-1}} \varphi - c_{2\mu-1} \int_{k_{2\mu}} \varphi), \qquad c_j = \int_{k_j} dg, \qquad j = 1, \ldots, g.$$

Note that all numbers $c_j (j = 1, \ldots, 2g)$ are real.

Let $h = \sum_{k=1}^{m} \alpha_k p_k, \sum_{k=1}^{m} \alpha_k = 0$. Fix an arbitrary point $p_0 \in M$ and connect this point with the points p_k by arcs $\breve{p_0 p_k} (k = 1, \ldots, m)$. We denote the obtained chain by $h_1, \partial h_1 = h, M - h_1 = M_1$.

Because of (12.21), we have in the domains $U_i \cap Uj$

$$T_i - T_j = \frac{1}{2\pi i} \ln(h_i/h_j).$$

Hence the function

$$(12.26)\qquad S(p) = -\frac{1}{2\pi i} \ln h_i(p) + T_i(p)$$

is single-valued on M_1. This involves the relation

$$\int_{M_1} \bar{\partial} t_i \wedge \varphi = \int_{M_1} \bar{\partial} S(p) \wedge \varphi = \int_{\partial M_1} S(p) d\Phi,$$

where Φ is some branch of the Abelian integral of the first kind corresponding to $\varphi, d\Phi = \varphi$. The boundary of M_1 consists of arcs of 1-chain h_1 passing in the following order (see Figure 8, p.25),

$$\partial M_1 = \breve{p_0 p_1} - \breve{p_0 p_1} + \cdots + \breve{p_0 p_m} - \breve{p_0 p_m}.$$

If the point p passes around the point p_k, the function $S(p)$ takes an increment $\Delta S(p) = -\alpha_k$, and hence

$$(12.27)\quad \begin{aligned} \int_{\partial M_1} S(p) d\Phi &= \sum_{k=1}^{m} \int_{p_0}^{p_k} \Delta S(p) d\Phi(p) = -\sum_{k=1}^{m} \alpha_k \int_{p_0}^{p_k} d\Phi \\ &= -\sum_{k=1}^{m} \alpha_k \Phi(p_k) + \Phi(p_0) \sum_{k=1}^{m} \alpha_k. \end{aligned}$$

The later addendum is equal to zero according to the condition $\deg h = 0$.

Comparing Eq. (12.25) and Eq. (12.27) we obtain the relation

$$(12.28)\quad \begin{aligned} \sum_{k=1}^{m} \alpha_k \Phi(p_k) &= \sum_{\mu=1}^{g} (c_{2\mu} \int_{k_{2\mu-1}} d\Phi - c_{2\mu-1} \int_{k_{2\mu}} d\Phi), \\ d\Phi &= \varphi \in H^0(Q^1), \end{aligned}$$

which is valid for any divisor of zero degree.

In order that $e^{*-1}(\delta^* h) \in i^* H^1(J)$, it is necessary and sufficient that all periods $c_j (j = 1, \ldots, 2g)$ of the form dg be integers. In fact, let the cocycle $\{\frac{1}{2\pi i} \ln \frac{h_k}{h_j}\}$ representing the cohomology class $e^{*-1}(\delta^* h)$ be an integer. Then

$$\frac{1}{2\pi i} \ln(h_k/h_j) = m_k - m_j + m_{kj}, \tag{12.29}$$

where $\{m_j\} \in Cn^0(Q)$ and $\{m_{ij}\} \in Z^1(J)$. In this case, the function

$$s(p) = \exp 2\pi i [t_j(p) - m_j(p)], \qquad p \in U_j \tag{12.30}$$

is defined on all surfaces M, since

$$\begin{aligned} s_k(p) s_j^{-1}(p) &= \exp 2\pi i \{t_k(p) - m_k(p) - t_j(p) + m_j(p)\} \\ &= \exp 2\pi i m_{jk} = 1, \qquad p \in U_j \cap U_k. \end{aligned}$$

This means that

$$g(p) = \frac{1}{2\pi i} \ln s(p), \tag{12.31}$$

and the numbers

$$c_j = \frac{1}{2\pi i} \int\limits_{k_j} d \ln s(p), \qquad j = 1, \ldots, 2g$$

are integers.

Conversely, if the numbers c_j are integers, the function $s(p) = \exp 2\pi i g(p)$ is single-valued on M. Then

$$t_j(p) = \frac{1}{2\pi i} \ln s(p) + m_j(p), \qquad p \in U_j, \qquad j \in I, \tag{12.32}$$

where the functions $m_j(p)$ are analytic in U_j, and

$$\frac{1}{2\pi i} \ln \frac{h_k(p)}{h_j(p)} = t_k(p) - t_j(p) = m_k(p) - m_j(p) + m_{kj}, \tag{12.33}$$

where $\{m_{kj}\}$ are integers determined by the difference of branches of the function $\frac{1}{2\pi i} \ln s(p)$ in the domains U_k and U_j,

$$m_{kj} = \left(\frac{1}{2\pi i} \ln s(p)\right)_k - \left(\frac{1}{2\pi i} \ln s(p)\right)_j.$$

Equation (12.28) is valid for any Abelian integral of the first kind. This involves the following theorem.

THEOREM 12.2 (ABEL). *For the second Cousin problem to be solvable for the divisor $h = \sum_{k=1}^{m} \alpha_k p_k, \deg h = 0$, it is necessary and sufficient that the comparison system*

$$\sum_{k=1}^{m} \alpha_k w_\mu(p_k) \equiv 0 \quad (\text{mod } periods\ of\ w_\mu), \qquad \mu = 1, \dots, g \tag{12.34}$$

be valid. Here $w_\mu (\mu = 1, \dots, g)$ is a complete system of Abelian integrals of the first kind.

In (12.34) we have used the symbol of the comparing modulo periods (see §6).

The necessity of (12.34) was proved above. In order to prove the sufficiency, note that Eq. (12.28) defines the real numbers $c_j (j = 1, \dots, 2g)$ uniquely. In fact, separating the real and imaginary parts in (12.28), we obtain that the vector $(c_2, -c_1, \dots, c_{2g}, -c_{2g-1})$ is a solution of the system of linear equations whose determinant

$$\begin{pmatrix} \operatorname{Re} \int\limits_{k_1} dw_1 & \dots & \operatorname{Re} \int\limits_{k_{2g}} dw_1 \\ \vdots & \dots & \vdots \\ \operatorname{Re} \int\limits_{k_1} dw_g & \dots & \operatorname{Re} \int\limits_{k_{2g}} dw_g \\ \operatorname{Im} \int\limits_{k_1} dw_1 & \dots & \operatorname{Im} \int\limits_{k_{2g}} dw_1 \\ \vdots & \dots & \vdots \\ \operatorname{Im} \int\limits_{k_1} dw_g & \dots & \operatorname{Im} \int\limits_{k_{2g}} dw_g \end{pmatrix} \tag{12.35}$$

is different from zero. In the opposite case, a zero combination of rows generates a differential of the first kind such that

$$\operatorname{Re} \int\limits_{k_1} dw = 0, \dots, \qquad \operatorname{Re} \int\limits_{k_{2g}} dw = 0.$$

This implies $dw = 0$. Therefore, the periods of the form dg are integers, and consequently, the class $e^{*-1}(\delta^* h) \in i^* H^1(J)$ is an integer.

C. *Classification of Complex Line Bundles*

In this section we shall classify the complex line bundles with $c(B) = 0$. From (12.2) it follows that $\operatorname{Ker} c = \operatorname{Ker} = \operatorname{Im} e^*$, i.e.,

$$\operatorname{Ker} c \cong H^1(Q)/i^* H^1(J). \tag{12.36}$$

Calculate $H^1(Q)$. The exact sequence

$$0 \to C \xrightarrow{i_0} Q \xrightarrow{d} Q^1 \to 0, \tag{12.37}$$

where i_0 is the embedding, involves the exact cohomology sequence

(12.38) $$\dots \xrightarrow{\delta_0^*} H^1(C) \xrightarrow{i_0^*} H^1(Q) \xrightarrow{d^*} H^1(Q^1) \xrightarrow{\delta_1^*} H^2(C) \to 0.$$

Taking into account that $H^2(C) \cong C$ and $H^1(Q^1) = H^0(Q^0) = C$ (see §10C), we find that the image of the homomorphism d^* is zero, i.e., the homomorphism $i_0^* : H^1(C) \to H^1(Q)$ is an epimorphism, and hence

(12.39) $$H^1(Q) = H^1(C)/\delta_0^* H^0(Q^1).$$

Then, as follows from (12.36),

(12.40) $$\operatorname{Ker} e = [H^1(C)/\delta_0^* H^0(Q^1)]/i^* H^1(J),$$

i.e.,

(12.41) $$\operatorname{Ker} c = H^1(C)/[\delta_0^* H^0(Q^1) + H^1(J)].$$

Reverting to Eq. (12.36), note that because of the Serre theorem, $H^1(Q) = H^0(Q^1)$, and consequently, the group $H^1(Q)$ possesses the structure of C^g. Every element of the group $H^1(Q)$ defines the functional on $H^0(Q^1)$. On the elements of the basis $dw_1, \dots, dw_g$ this functional takes the values

(12.42) $$l_\mu = \int_M \bar\partial t_i \wedge dw_\mu, \qquad \mu = 1, \dots, g.$$

On the group $H^1(J)$ the values l_μ are equal to zero modulo periods of w_μ. Therefore, every point of $\operatorname{Ker} c$ is determined by the g complex numbers $l_1, \dots, l_g$ modulo periods of Abelian integrals of the first kind. Consequently, the group $\operatorname{Ker} c$ has the structure of the g-dimensional complex torus (the Picard group). This matter is studied in detail in the books [10], [17], and [18].

D. *Solvability of the Riemann Problem,* $\kappa = 0$

Consider the solvability conditions of the Riemann problem for $0 \leqslant \kappa < g$.

First, let $\kappa = 0$. We suppose that for every connected component of the contour L, $L = \sum_{j=1}^m L_j$,

$$\kappa_j = \operatorname{ind}_{L_j} G = 0, \qquad j = 1, \dots, m.$$

Denote by M_L the multiplicative sheaf of germs of holomorphic functions in the domains $T^\pm$, continuous up to the boundary and different from zero on L, and consider the exact sequence

(12.43) $$0 \to Q^* \xrightarrow{i} M_L^* \to \theta_L \to 0.$$

The third term of (12.43) $\theta_L = M_L^*/Q^*$. By analogy with the sheaf of germs of divisors, the sheaf θ_L is called the sheaf of germs of boundary problems. Every

Riemann problem corresponds to some element of the group $H^0(\theta_L)$ of sections of θ_L. We denote this section by h. For the solvability of the Riemann problem, it is necessary and sufficient that in the sequence

$$0 \to H^0(Q^*) \xrightarrow{i^*} H^0(M_L^*) \xrightarrow{\pi^*} H^0(\theta_L) \xrightarrow{\delta^*} H^1(Q^*) \tag{12.44}$$

the element $h \in H^0(\theta_L)$ has the preimage $\pi^{*-1}(h)$, i.e., $h \in \pi^* H^0(M_L^*)$ According to the exactness of (12.44), this is equivalent to the condition $\delta^* h = 0$. We shall calculate this class. It is determined by the cocycle

$$h_{ij} = \frac{\varphi_i^{\pm}(p)}{\varphi_j^{\pm}(p)}, \tag{12.45}$$

where $\{\varphi_i^{\pm}\}$ was defined in §11C.

Due to the condition that $\operatorname{ind}_L G = c(B_G) = 0$, the class $\delta^* h$ has a preimage for the homomorphism e^* (12.2). The class $e^{*-1}(\delta^* h)$ is defined by the cocycle

$$\left\{\frac{1}{2\pi i} \ln \frac{\varphi_i}{\varphi_j}\right\} \in Z^1(Q). \tag{12.46}$$

The isomorphism $H^1(Q) \to \Gamma(A^{0,1})/\bar\partial\Gamma(A^0)$ transforms the class corresponding to the cocycle (12.46) onto the class containing the form $\bar\partial t_i$,

$$t_i = \frac{1}{2\pi i} \sum_{k \in I} \alpha_k \ln \frac{\varphi_i}{\varphi_k}. \tag{12.47}$$

According to the Serre theorem, this class corresponds to the functional on the group $H^0(Q^1)$ defined by the relation

$$(\bar\partial t_i, dw) = \int_M \bar\partial t_i \wedge dw, \qquad dw \in H^0(Q^1). \tag{12.48}$$

For this integral, the relation on (12.35) is valid. On the other hand, the function

$$S(p) = \frac{1}{2\pi i} \ln \varphi_j + t_j(p), \qquad p \in U_j \tag{12.49}$$

is continuous on ∂U_j (analogous to (12.26)). On the surface $\widetilde{M}$, obtained from M by cutting along the contour L, we have

$$\int_M \bar\partial t_i \wedge dw = \int_{\widetilde{M}} \bar\partial S(p) \wedge dw = \int_{\partial\widetilde{M}} S(p)\,dw.$$

The boundary of $\widetilde{M}$ is the contour L passing twice in opposite directions

$$\begin{aligned} \int_{\partial\widetilde{M}} S(p)\,dw &= \int_L S^+(p)\,dw - \int_L S^-(p)\,dw \\ &= \frac{1}{2\pi i} \int_L (\ln \varphi_i^+ - \ln \varphi_i^-)\,dw = \frac{1}{2\pi i} \int_L \ln G\,dw. \end{aligned} \tag{12.50}$$

Comparing (12.25) and (12.50), we obtain the relation

$$\frac{1}{2\pi i}\int_L \ln G\,dw = \sum_{\mu=1}^{g}\Big(c_{2\mu}\int_{k_{2\mu-1}} dw - c_{2\mu-1}\int_{k_{2\mu}} dw\Big), \tag{12.51}$$

which is valid for any Abelian differential of the first kind. As was shown in §12B for $e^{*-1}(\delta^* h) \in H^1(J)$, it is necessary and sufficient that c_j be integers.

If $\kappa = 0$ but $\kappa_j \neq 0 (j = 1, \ldots, m)$, we use the function $\pi(p)$ holomorphic in T^+, different from zero and such that $\operatorname{ind}_{L_j} \pi = -\kappa_j (j = 1, \ldots, m)$. To construct such a function, we use the surface $\widetilde{T}$, obtained from T^+ by the following operation. The contours $L_1, \ldots, L_n$ are sealed up by simply connected domains $S_j (j = 1, \ldots, n)$. In every domain S_j a point p_j is chosen, and the function $\pi(q)$ is determined in the domain T bounded by the contour L_0 by the divisor $(\pi) = -\sum \kappa_j p_j$. Assume $\widehat{G}(p) = G(p)\pi(p)$. It is clear that $\operatorname{ind}_{L_j} G(p) = 0, j = 1, \ldots, m$.

THEOREM 12.4. *In order that the Riemann problem of zero index be solvable, it necessary and sufficient that*

$$\int_L \ln \hat{G}(p)\,dw_\mu(p) \equiv 0 \quad (\text{mod } periods\ of\ w_\mu), \qquad \mu = 1, \ldots, g. \tag{12.52}$$

E. *Solvability of the Riemann Problem,* $0 < \kappa < g$

Let the Riemann problem have a holomorphic solution. Then the line bundle B_G determines a positive divisor $h = \sum \alpha_k p_k \geqslant 0, h = h^+ + h^-, h^\pm = h \cap T^\pm, \deg h = \operatorname{ind}_L G$.[3] The solution of the boundary problem is represented by

$$F^\pm(p) = h^\pm(p) f^\pm(p), \tag{12.53}$$

and $f^\pm$ is a solution of the problem

$$f^+(p) = G(p)\frac{h^-(p)}{h^+(p)} f^-(p) \tag{12.54}$$

of the zero index. Here $h^\pm(q)$ are functions holomorphic in $T^\pm$ whose divisors are $h^\pm$.

On account of (12.52) we have

$$\frac{1}{2\pi i}\int_L \ln\left[\widehat{G}(p)\frac{h^-(p)}{h^+(p)}\right] dw_\mu \equiv 0 \quad (\text{mod periods of } w_\mu). \tag{12.55}$$

---

[3]For simplicity, we suppose that zeros are absent on L. If a point $p_k \in L$, the following considerations must be slightly modified, assuming that this zero belongs to both domains $T^\pm$. In this case, we may obtain the same result (12.56).

Fix on every connected component $L_j (j = 0, \ldots, m)$ of the contour L an arbitrary point p_j and fix the single-valued branch of every function $f(p)$ on L_j by the condition $f(p_j) = 0$. It is clear that branches of multivalued functions are discontinuous on L_j and have jumps at the point p_j. Rewrite (12.55) as

$$\frac{1}{2\pi i}\int\limits_L \ln \widehat{G} dw_\mu + \frac{1}{2\pi i}\int\limits_L \ln \frac{h^-(p)}{h^+(p)} dw_\mu \equiv 0 \quad (\text{mod periods of } w_\mu).$$

The second integral can be easily calculated. In fact, if an analytic function $f_q(p)$ in T^+ has a single pole of the first order at the point q, then by drawing a cut from the point $p_0 \in L$ to q, we obtain

$$\frac{1}{2\pi i}\int\limits_L \ln f_q(p) dw_\mu + \frac{1}{2\pi i}\int_{p_0}^{q} \ln f_q dw_\mu - \frac{1}{2\pi i}\int_{p_0}^{q} \ln f_q dw_\mu = 0,$$

where the second and the third integrals are taken over different sides of the cut. When the point p passes around the point q, the function $\ln f_q(p)$ takes the increment $2\pi i$. Thus we obtain

$$\frac{1}{2\pi i}\int\limits_L \ln f_q dw = \frac{1}{2\pi i}\int_q^{p_0} 2\pi i dw = w(q) - w(p_0).$$

We have

$$\frac{1}{2\pi i}\int\limits_L \ln h^+ dw = \sum_{p_k \in T^+} \alpha_k w(p_k) - \sum_{p_k \in T^+} \alpha_k w(p_0).$$

The integral of $\ln h^-$ is calculated in an analogous manner. If the Riemann problem is solvable, then

$$(12.56)\quad \frac{1}{2\pi i}\int\limits_L \ln \widehat{G} dw_\mu - \sum_k \alpha_k \int_{p_0}^{p_k} dw_\mu \equiv 0 \ (\text{mod periods of } w_\mu), \qquad \mu = 1, \ldots, g.$$

Conversely, if for some divisor $\sum \alpha_k p_k$ the condition (12.56) is valid, then the Riemann problem is solvable, and the zeros of the solution are determined by h. Therefore, we obtain the following theorem.

THEOREM 12.5. *In order that the Riemann problem be solvable, it is necessary and sufficient that the inversion problem*

$$(12.57)\qquad \sum_{k=1}^{\kappa} w_\mu(p_k) = c_\mu \ (\text{mod periods of } w_\mu),$$

$$(12.58)\qquad c_\mu = \frac{1}{2\pi i}\int\limits_L \ln G dw_\mu, \quad \mu = 1, \ldots, g$$

be solvable.

F. *The Nonhomogeneous Riemann Problem*

In conclusion, we obtain the conditions of solvability for the Riemann problem

$$F^+(p) = G(p)F^-(p) + g(p) \tag{12.59}$$

(cf. §5). Consider the sheaf $Q(B_G)$ of germs of holomorphic sections of the bundle B_G. Let M_L be the sheaf of germs of holomorphic functions on $M \backslash L$ and continue up to the boundary (to the right and to the left of L).

Consider the exact sequence

$$0 \to Q(B) \xrightarrow{i} M_L \xrightarrow{\pi} \eta \to 0, \tag{12.60}$$

where $\eta = M_L/Q(B_G)$. We obtain the exact sequence

$$H^0(M_L) \xrightarrow{\pi^*} H^0(\eta) \xrightarrow{\delta^*} H^1(Q(B_G)). \tag{12.61}$$

Every problem (12.59) corresponds to some element g of the group $H^0(\eta)$. This element is determied by the cochain $\{F_i\}$ of germs of the solution of the problem (12.59). The solvability of the problem (12.59) means that the element $g \in H^0(\eta)$ has the preimage $\pi^{*-1}(g)$. Hence, the necessary and sufficient condition of the solvability of problem (12.59) is $\delta^* g = 0$. The class $\delta^* g$ is determined by the cocycle $\{F_{ij}\}, F_{ij} = F_i - F_j$ in $U_i \cap U_j$. The functions F_{ij} satisfy the boundary condition (11.14) on L.

According to the Serre theorem, every element of $H^1(Q(B_G))$ determines the linear functional on the group $H^0(Q^1(-B_G)) = H^0(Q(-B_G + K))$ of solutions of the problem (11.15) dual to (11.14). This functional has the form

(12.62)

$$\begin{aligned}\int_M \bar\partial F_i \wedge \omega_i &= \int_{T^+} \bar\partial F_i \wedge \omega_i + \int_{T^-} \bar\partial F_i \wedge \omega_i \\ &= \int_{\partial T^+} F_i\omega_i + \int_{\partial T^-} F_i\omega_i = \int_L F_i^+\omega_i^+ - \int_L F_i^-\omega_i^- \\ &= \int_L F_i^+\omega_i^+ - \int_L \frac{1}{G}(F_i^+ - g)G\omega_i^+ = \int_L g\omega_i^+, \{\omega_i\} \in H^0(Q^1(-B_G)).\end{aligned}$$

The condition $\delta^* g = 0$ is equivalent to the requirement that the functional (12.62) be trivial on the group $H^0(Q^1(-B_G))$. Therefore we obtain the following theorem.

THEOREM 12.6. *In order that the problem (12.59) be solvable, it is necessary and sufficient that for any solution $\omega^\pm$ of the dual problem (11.15), the condition*

$$\int_L g\omega^+ = 0 \tag{12.63}$$

be valid (cf. Theorem 5.5).

CHAPTER 3

THE RIEMANN BOUNDARY PROBLEMS FOR VECTORS ON COMPACT RIEMANN SURFACES

§13 The Riemann Boundary Problem for Vector Functions

A. *The Riemann Problem and Complex Vector Bundles*

The theory of complex vector bundles on Riemann surfaces covers a wide area rich in ideas and results, and is far from being complete (see, for example, [7], [17], [34]). Grothendieck [a] and Röhrl [c] had established the connection between vector bundles and the Riemann problem. In this chapter we shall expound the author's results of a general solution of the Riemann problem and, as a corollary, obtain the Riemann–Roch theorem for vector bundles on a Riemann surface.[1] We shall examine the problem

$$F^{+}(p) = G(p)F^{-}(p),$$

where $G(p)$ is a Hölder continuous $n \times n$-matrix which is nondegenerate on L, and $F^{\pm}(q)$ are analytical vector-columns or $n \times n$-matrices. We shall consider the holomorphic solution and solutions which are multiples of a given divisor. All notes concerning the statement of the problem in §5 remain valid. Note that a vector-function has a zero at some point if all its components have zero. The order of zero is defined as the minimum of orders of all components. A vector-function has a pole if at least one component of this vector has a pole. An order of pole is defined as the maximum of the pole order of all components.

First of all, as in §11, we shall establish a correspondence between the Riemann boundary problem and complex vector bundles.

---

[1]See also Appendix 2,3.

Consider a covering $\{U_i\}$ of the surface M and groups $M(U_i)$ defined as follows. If $U_i \cap L = 0$, the group $M(U_i)$ is the group of holomorphic vector-functions in U_i. If $U_i \cap L \neq 0$ and $U_i = T_i^+ \cup T_i^-$, we assume that $M(U_i)$ is the group of holomorphic vector-functions in $T^\pm$ which is continuous up to the $L \cap U_i$ and satisfies the condition (13.1) on $L \cap U_i$. The group $M(U_i)$ defines the sheaf Q_G of germs of holomorphic vector solutions of the problem (13.1). The vector bundle B_G with the fibre C^n and the structure group $GL(n, C)$ is determined by the transition matrices

$$\begin{aligned} h_{ij}(p) &= (F_i^+(p))^{-1}F_j^+(p) = (F_i^-(p))^{-1}F_j^-(p), \qquad p \in U_i \cap U_j, \\ &\qquad U_i \cap U_j \cap L \neq 0, \\ h_{ij}(p) &= E, \\ &\qquad U_i \cap U_j \cap L = 0, \end{aligned} \tag{13.2}$$

E is the unit matrix, and F_i are nondegenerate holomorphic $n \times n$-matrices in U_i satisfying the condition (13.1).

It is clear that the sheaf Q_G coincides with the sheaf of germs of holomorphic sections of the bundles $B_G, Q_G = Q(B_G)$. Every vector bundle B with fibre C^n may be associated with the so-called principal bundle $\widehat{B}$ with the same transition matrices and a fibre $GL(n, C)$. It is also clear that the sheaf of germs $Q(B_G)$ is determined by degenerate solutions of the problem (13.1) in the domains U_i.

THEOREM 13.1. *If B is a vector bundle over M with a fibre C^n and a structure group $GL(n, C)$, then for the given contour L there exists a degenerate $n \times n$-matrix $G(p)$, such that B is equivalent to the bundle B_G.*

In §14 it will be shown that in the domains $T^\pm$, the bundle B is trivial, i.e., is represented in the form $h_{ij} = h_j^{-1}h_i$, where $h_i(p)$ is an analytic degenerate matrix in U_i. If the contour L separates the domain U_i into two domains $U_i^\pm$, in every domain there are matrices $h_{ij}^\pm$ defined by,

$$h_{ij} = \begin{cases} (h_j^+)^{-1}h_i^+, & \text{in } U^+ \\ (h_j^-)^{-1}h_i^-, & \text{in } U^-. \end{cases}$$

Then we can define the matrix on L

$$G(p) = h_i^+(p)(h_i^-(p))^{-1} = h_j^+(p)(h_j^-(p))^{-1}, \qquad p \in L, \tag{13.3}$$

independent of a choice of a covering domain containing the point p.

A fundamental system of sections of a bundle B with fibre C^n is a set of n meromorphic sections over M such that their determinant is not identically equal to zero, i.e., a degenerate identically zero section of the principal bundle B_G.

A fundamental system of sections of the bundle B_G determines the fundamental system of solutions of the problem (13.1).

THEOREM 13.2. *A bundle B with fibre C^n and the structure group $GL(n, C)$ possesses a fundamental system of sections.*

Let h be some divisor and $\widetilde{B}_h$ be the direct sum of the complex line bundles B_h,

$$\widetilde{B}_h = \bigoplus_{i=1}^{n} B_h, \qquad h = \sum_k \alpha_k p_k.$$

Consider the exact sequence

$$0 \to Q(B) \to Q(B - \widetilde{B}_h) \to h \times C \to 0. \tag{13.4}$$

Here $h \times C$ is the sheaf whose fibre is different from zero only over points of the vector-divisor h. The fibre over point p_k is the direct sum $\bigoplus_{j=1}^{n} C^{\alpha_k}$. We have the exact sequence

$$\begin{aligned} 0 \to H^0(Q(B)) \to H^0(Q(B - \widetilde{B}_h)) \to H^0(h \times C) \\ \to H^1(Q(B)) \to H^1(Q(B - \widetilde{B}_h)) \to 0, \end{aligned}$$

since $H^1(h \times C) = 0$. It follows from the fact that for sufficiently refined coverings, no intersections $U_i \cap U_j$ contain any points p_k. This means that

$$\begin{aligned} n \deg h = \dim H^0(h \times C) = [\dim H^0(Q(B - \widetilde{B}_h) - \dim H^0(Q(B))] \\ + [\dim H^1(Q(B)) - \dim H^1(Q(B - \widetilde{B}_h))], \end{aligned} \tag{13.5}$$

and hence

$$\begin{aligned} \dim H^0(Q(B - \widetilde{B}_h)) = n \deg h + \dim H^0(Q(B)) \\ + \dim H^1(Q(B - B_h)) - \dim H^1(Q(B)). \end{aligned} \tag{13.6}$$

Therefore, $\dim H^0(Q(B - B_h)) \geqslant n$ for

$$n \deg h \geqslant \dim H^1(Q(B)) + n. \tag{13.7}$$

Holomorphic sections of the bundle $B - \widetilde{B}_h$ evidently correspond to meromorphic sections of B which are multiples of the divisor $-h$, $h = \bigoplus h_j$. We obtain the following theorem.

THEOREM 13.2′. *The problem (13.1) possesses a fundamental system of meromorphic solutions.*

Denote a fundamental system of solutions by $X(X_1, \ldots, X_n)$; the vector solutions X_i are determined by the divisors g_i, $(X_i) = g_i$, and the determinant divisor is

$$(\det X) = \sum_{i=1}^{n} g_i + g_0, \qquad g_0 \geqslant 0. \tag{13.8}$$

A fundamental system is called canonical if $g_0 = 0$, i.e., if an order of $\det X$ is equal to the sum of orders of columns at every point.

THEOREM 13.3. *In order for the problem (13.1) to be a canonical system of solutions meromorphic on M, it is necessary and sufficient that the bundle B_G be equivalent to a direct sum of complex line bundles.* (Rodin [p])

Indeed, let B_G be decomposed into a direct sum of line bundles. Then

$$h_{ij} = Z_i \operatorname{diag}(g_{1,ij}, \ldots, g_{n,ij}) Z_j^{-1}, \qquad p \in U_i \cap U_j, \tag{13.9}$$

where $\{g_{a,ij}\}(a = 1, \ldots, n)$ are cocycles corresponding to linear components of B_G, and Z_i are nondegenerate holomorphic matrices in U_i. Taking into account Eq. (13.2) and by replacing cocycles $\{g_{a,ij}\}$ with transition functions of corresponding divisors $\{g_{a,ij}\} = \{g_{ai} g_{aj}^{-1}\}$, we obtain the canonical system

$$F_i^{\pm} Z_i \operatorname{diag}(g_{1i}, \ldots, g_{ni}) = F_j^{\pm} Z_j \operatorname{diag}(g_{1j}, \ldots, g_{nj}) = X^{\pm}. \tag{13.10}$$

Conversely, if $X^{\pm}$ is a canonical solution whose columns are determined by the divisors $g_a = g_{ai}$, we assume that

$$Z_i = (F_i^{\pm})^{-1} X^{\pm} \operatorname{diag}(g_{1i}^{-1}, \ldots, g_{ni}^{-1}).$$

Then the transition matrices of the bundle B_G are represented by the form (13.9).

Note that our definition of a canonical system is wider than the one of a sphere [27] where all zeros and poles of the solution are assumed to be at ∞. Theorem 13.3 is a very convenient tool to verify the decomposition of the vector bundle in the direct sum of line bundles. For an easy calculation, we consider only the case $n = 2$.

Examine the problem

$$F^+ = GF^-, \qquad G = \begin{pmatrix} g_{11} & g_{12} \\ g_{21} & g_{22} \end{pmatrix}. \tag{13.11}$$

Let $\begin{pmatrix} \widehat{F}_1^{\pm} \\ \widehat{F}_2^{\pm} \end{pmatrix}$ be a specific solution of (13.11). We shall search for a solution of the form

$$F^{\pm} = \begin{pmatrix} \widehat{F}_1^{\pm} & 1 \\ \widehat{F}_2^{\pm} & 0 \end{pmatrix} \begin{pmatrix} Y_1^{\pm} \\ Y_2^{\pm} \end{pmatrix} = \begin{pmatrix} \widehat{F}_1^{\pm} Y_1^{\pm} + Y_2^{\pm} \\ \widehat{F}_2^{\pm} Y_1^{\pm} \end{pmatrix}. \tag{13.12}$$

Then

$$Y^+ = \begin{pmatrix} 1 & c \\ 0 & b \end{pmatrix} Y^-, \qquad Y^{\pm} = \begin{pmatrix} Y_1^{\pm} \\ Y_2^{\pm} \end{pmatrix},$$

$$b = g_{11} - \frac{\widehat{F}_1^+}{\widehat{F}_2^+} g_{21}, \qquad c = \frac{g_{21}}{\widehat{F}_2^+}. \tag{13.13}$$

Therefore, the problem of the decomposition of B_G is reduced to the problem of the existence of a canonical solution of the problem (13.13).

THEOREM 13.4. *For problem (13.13) to have a canonical solution, it is necessary and sufficient that*

$$\int_L c\omega^- = 0 \tag{13.14}$$

for every holomorphic differential satisfying the boundary condition

$$\omega^+ = b^{-1}\omega^-. \tag{13.15}$$

Problem (13.13) may be rewritten as

$$Y_1^+ = Y_1^- + cY_2^-, \tag{13.16}$$

$$Y_2^+ = bY_2^-. \tag{13.17}$$

We have one specific solution $Y_1^\pm = 1$, $Y_2^\pm = 0$. We shall look for the second column of a canonical solution.

Let the condition (13.14) be valid. Then the problem

$$c = c^- - bc^+ \tag{13.18}$$

is solvable, $c^\pm$ are holomorphic, and (13.16) may be rewritten as

$$Y_1^+ + c^+Y_2^+ = Y_1^- + c^-Y_2^-. \tag{13.19}$$

The functions $Y_2^\pm$ can be obtained from (13.17). We have from (13.19)

$$Y_1^\pm(q) = -c^\pm(q)Y_2^\pm(q) + F(q), \tag{13.20}$$

where $F(q)$ is an arbitrary analytic function on M, such that $(F) \geqslant (Y_2^\pm)$. Then $(Y_1^\pm) \geqslant (Y_2^\pm)$, and the matrix

$$Y = \begin{pmatrix} Y_1^\pm & 1 \\ Y_2^\pm & 0 \end{pmatrix} \tag{13.21}$$

is the canonical solution of (13.13), since $\det Y = -Y_2^\pm$, and the divisor $(\det Y)$ coincides with the divisor of $\begin{pmatrix} Y_1^\pm \\ Y_2^\pm \end{pmatrix}$.

Conversely, if (13.21) is a canonical solution of (13.13), then $(Y_1^\pm) \geqslant (Y_2^\pm)$ and $Y_1^\pm/Y_2^\pm$ are holomorphic in $T^\pm$. Assuming in (13.16) that $c^\pm = Y_1^\pm/Y_2^\pm$, we conclude that problem (13.18) is solvable and hence the condition (13.14) is valid.

B. *The General Solution of the Riemann Problem*

Let $X = (X_1, \dots, X_n)$ be a fundamental solution of the problem (13.1), and the divisors of the columns and of the determinant of the solution are related by Eq.(13.8),

$$(\det X) = \sum_{a=1}^{n} g_a + g_0.$$

We suppose that the divisor g_0 contains no multiple points (for the general case, see Rodin [p,q]). We suppose also that the divisor $g_0 = \sum_{k=1}^{m} p_k$ (all p_k are different) has no common points with the divisors g_a, $g_a \cap g_0 = \emptyset (a = 1, \dots, n)$.

The general solution of (13.1) is represented by

$$F(q) = \sum_{a=1}^{n} X_a(q) P_a(q), \tag{13.22}$$

where $P_a(q)$ are rational functions on M which are multiples of the divisors $-(g_a + g_0)$. Poles of $P_a(q)$ at the points of g_0 must be compensated by the linear dependence of the functions $X_j(q)$ at these points.

This means that if for some fixed local coordinates $z(p_k)$ we have for the b-th component X_{ab} of the vectors X_a,

$$P_a(z(q)) = \frac{c_{ka}}{z(q) - z(p_k)} \qquad + \text{regular terms}, \tag{13.23}$$
$$a = 1, \dots, n, \qquad k = 1, \dots, m,$$

then[2]

$$\sum_{a=1}^{n} c_{ka} X_{ab}(p_k) = 0, \qquad k = 1, \dots, m, \qquad b = 1, \dots, n. \tag{13.24}$$

Consider the exact sequence

$$0 \longrightarrow Q(B_{g_1 \dots g_n}) \stackrel{i}{\longrightarrow} Q(B_{g_1 + g_0 \dots g_n + g_0}) \stackrel{\pi}{\longrightarrow} Q_{g_0 \dots g_0} \longrightarrow 0. \tag{13.25}$$

Here $B_{g_1, \dots, g_n} = \bigoplus_a B_{g_a}$ is the direct sum of the line bundles corresponding to the divisors g_a, and $Q_{g_0, \dots, g_0}$ is the quotient-sheaf which is isomorphic to the sheaf of germs of n-tuple principal parts at the points of the divisor g_0.

Let $S_{g_0} \subset Q_{g_0}, \dots, g_0$ be the subsheaf of the principal parts satisfying the condition (13.24).

[2]The value c_{ka} is similar to Tyurin's parameters [34] and Krichever & Novikov [a].

THEOREM 13.5.

$$\dim H^0(S_{g_0}) = \deg g_0. \tag{13.26}$$

It is sufficient to show that for any point $p_k \in g_0$

$$\dim H^0(S_{p_k}) = 1.$$

On account of Eq. (13.24),

$$\dim H^0(S_p) = \text{Corank}\,(X_{ab(P_k)}) = n - \text{rank}\,(X_{ab}(p_k)).$$

The value rank $(X_{ab}(p_k)) < n$, since Eq. (13.24) is solvable. If rank $(X_{ab}(p_k)) < n-1$, then X has at p_k the order of more than 1. Hence rank $(X_{ab}(p_k)) = n-1$, and Eq. (13.26) is valid.

The sequence (13.25) involves the diagram

(13.27)
$$0 \longrightarrow H^0(Q(B_{g_1\ldots g_n})) \xrightarrow{i^*} H^0(Q(B_{g_1+g_0\ldots g_n+g_0}))$$

$$\begin{array}{ccccccc}
\xrightarrow{\pi^*} & H^0(Q_{g_0\ldots g_0}) & \xrightarrow{\delta^*} & H^1(Q(B_{g_1\ldots g_n})) & \xrightarrow{i^1} & H^1(Q(B_{g_1+g_0\ldots g_n+g_0})) \\
& \uparrow j^* & & \downarrow \text{duality} & & \downarrow \text{duality} \\
& H^0(S_{g_0}) & & H^0(Q^1(-B_{g_1\ldots g_n})) & \xleftarrow{\tilde{i}} & H^0(Q^1(-B_{g_1+g_0\ldots g_n+g_0}))
\end{array}$$

The linear space L_G of solutions of the problem (13.1) is isomorphic to

$$\pi^{*-1}j^*H^0(S_{g_0}).$$

In order to study this space, consider first the space

$$L_0 = \pi^{*-1}j^*H^0(S_{g_0})/i^*H^0(Q(B_{g_1\ldots g_n})).$$

It is isomorphic to the group

$$L \cong j^*H^0(S_{g_0}) \cap \text{Ker}\,\delta^*,$$

where δ^* is a coboundary operator in the sequence (13.27). Since j^* is an embedding, the group $j^*H^0(S_{g_0})$ is described by Theorem 13.5.

In order to describe Ker δ^*, calculate the group $\delta^*j^*H^0(S_{g_0})$. Consider the larger group $\delta^*H^0(Q_{g_0,\ldots,g_0})$. Due to the Serre theorem, this group can be considered as a subgroup of the group of linear functionals on the group $H^0(Q^1(-B_{g_1,\ldots,g_n}))$. Because $\delta^*H^0(Q_{g_0,\ldots,g_0}) = \text{Ker}\,i^1$, every element of this group corresponds to the cocycle belonging to the zero cohomology class of the group $H^1(Q(B_{g_1+g_0,\ldots,g_n+g_0}))$, i.e., every cocycle $\{h_{ij}\}$ can be represented by

$$h_{ij} = h_i h_j^{-1}, \qquad \{h_i\} \in Cn^0(Q(B_{g_1+g_0,\ldots,g_n+g_0})).$$

This cocycle defines the functional on the group given by $H^0(Q^1(-B_{g_1,\dots,g_n}))$ (13.28)

$$< \{h_{ij}\}, \varphi >= \sum_M \operatorname{Res}(h_i, \varphi) = \sum_M \operatorname{Res} \sum_{a=1}^n h_{ia}\varphi_a = \sum_{p_k \in g_0} \sum_{a=1}^n c_{ak}\varphi_a(p_k).$$

Here $\varphi(\varphi_1, \dots, \varphi_n)$ is a vector-differential belonging to $H^0(Q^1(-B_{g_1,\dots,g_n}))$, $\langle \cdot, \cdot \rangle$ is the symbol of the inner product, and the coefficients c_{ak} are determined by Eq. (13.24). We can easily verify that the value of Eq. (13.28) depends only on the cohomology class of the cocycle $\{h_{ij}\}$, since for a cocycle cohomological to zero all $c_{ka} = 0$.

Therefore, the functional (13.28) is defined on the product of the groups

$$H^1(Q(B_{g_1,\dots,g_n}) \times H^0(Q^1(-B_{g_1,\dots,g_n})).$$

On the group $\delta^* H^0(Q_{g_0,\dots,g_0})$ the functional (13.28) vanishes for all vector-differentials whose components are multiples of the divisor g_0. We obtain the duality

$$\delta^* H^0(Q_{g_0\dots g_0}) \quad \text{dual to} \quad H^0(Q^1(-B_{g_1\dots g_n}))/\tilde{i}H^0(Q^1(-B_{g_1+g_0\dots g_n+g_0})).$$

The same result can be obtained from the right lower square of the diagram (13.27), since $\delta^* H^0(Q_{g_1,\dots,g_0}) = \operatorname{Ker} i^1$, and

$$H^1(Q^1(-B_{g_1\dots g_n}))/\tilde{i}H^0(Q^1(-B_{g_1+g_0\dots g_n+g_0})) = \operatorname{Coker} i^1.$$

Let us consider the subgroup of the group $\operatorname{Coker} i^1$ annihilated by the group $\delta^* j^* H^0(S_{g_0})$ and denote it by $H^0(Q^1_{g_0}(-B_{g_1,\dots,g_n}))$. We obtain, therefore, the following result.

By the definition

$$L_0 = L_G/i^* H^0(Q(B_{g_1\dots g_n})).$$

This group is isomorphic to the group $j^* H^0(S_{g_0}) \cap \operatorname{Ker} \delta^*$, where the group $j^* H^0(S_{g_0})$ was described by Theorem 13.4, and the group $\delta^* j^* H^0(S_{g_0})$ is dual to the group

$$H^0(Q^1(-B_{g_1\dots g_n}))/\tilde{i}H^0(Q^1_{g_0}(-B_{g_1\dots g_n})).$$

Theorem 13.5 involves the formula for the number of solutions of the problem (13.1),

$$\begin{aligned}
\dim L_G &= \dim H^0(Q(B_{g_1\dots g_n})) + \{\dim H^0(S_{g_0}) \\
&\qquad - \dim H^0(Q^1(-B_{g_1\dots g_n})) + \dim H^0(Q^1_{g_0}(-B_{g_1\dots g_n}))\} \\
&= \sum_{a=1}^n \dim H^0(Q(B_{g_a})) + \deg g_0 \\
&\qquad - \sum_{a=1}^n \dim H^0(Q^1(-B_{g_a})) + \dim H^0(Q^1_{g_0}(-B_{g_1\dots g_n})) \\
&= \sum_{a=1}^n \deg g_a + \deg g_0 - ng + n \\
&\qquad \dim H^0(Q^1_{g_0}(-B_{g_1\dots g_n})).
\end{aligned}$$

Therefore, we have proved the following theorem.

THEOREM 13.6.

$$\dim L_G = \operatorname{ind}_L \det G - ng + n + \dim H^0(Q^1_{g_0}(-B_{g_1 \ldots g_n})). \tag{13.29}$$

C. *The Conjugate Problem. The Riemann–Roch Theorem for Vector Bundles*

Consider the problem conjugate to (13.1)

$$f^+ = G^{*-1} f^- \tag{13.30}$$

for vector-differentials. Here G^* is the transposed matrix of G.

Let X be a fundamental system of solutions of the problem (13.1). Then[3]

$$G(p) = X^+(p)[X^-(p)]^{-1}.$$

From (13.30) we obtain

$$f^+(p) = [X^+(p)]^{*-1}[X^-(p)]^* f^-(p).$$

Therefore, the vector-differential

$$\omega = X^* f = \begin{cases} [X^+(q)]^* f^+(q), & q \in T^+, \\ [X^-(q)]^* f^-(q), & q \in T^- \end{cases} \tag{13.31}$$

is a meromorphic differential on M, continuous on L.

Denoting the a-th component of the vector ω by ω_a and the a-th column of the matrix $X^\pm$ by $X^\pm_a$, we obtain

$$\omega_a = < X^\pm_a, f >, \qquad a = 1, \ldots, n. \tag{13.32}$$

LEMMA. *The differential* ω *satisfies the condition*

$$L[\omega] = \sum_{p_k \in g_0} \sum_{a=1}^{n} c_{ka}\omega_a(p_k) = 0, \tag{13.33}$$

where the coefficients c_{ka} *are determined by the Eq. (13.24).*

Eq. (13.32) involves the relation

$$\omega_a(p_k) = < X_a(p_k), f(p_k) > .$$

Substituting for this relation in (13.33) and taking into account (13.24), we obtain

$$L(\omega) = \sum_{p_k \in g_0} \sum_{a=1}^{n} c_{ka} < X_a(p_k), f(p_k) > = \sum_{p_k \in g_0} < \sum_{a=1}^{n} c_{ka} X_a(p_k), f(p_k) > = 0.$$

---

[3]For simplicity we suppose that zeros and poles of $\det X$ do not belong to the contour L.

Eq. (13.31) implies that the vector-differential ω determines the cohomology class

$$\{\omega\} \in H^0(Q^1(-B_{g_1,\dots,g_n})).$$

From the lemma and Eq. (13.28) it follows that

$$\{\omega\} \in H^0(Q^1_{g_0}(-B_{g_1\dots g_n})). \tag{13.34}$$

By virtue of (13.31) the general solution of the problem (13.30) has the form

$$f^{\pm} = (X^{\pm})^{*-1}\omega. \tag{13.35}$$

THEOREM 13.7. *The group H_G of solutions of the problem (13.30) is isomorphic to the group*

$$H_G \cong H^0(Q^1_{g_0}(-B_{g_1\dots g_n})). \tag{13.36}$$

From (13.34) we obtain the embedding

$$i : H_G \to H^0(Q^1_{g_0}(-B_{g_1\dots g_n})), \qquad \operatorname{Ker} i = 0. \tag{13.37}$$

We must show that it is an epimorphism. The embedding (13.37) and Eq.(13.29) imply the inequality

$$\dim L_G - \dim H_G \geqslant \operatorname{in} d_L \det G - ng + n. \tag{13.38}$$

We have to show that (13.38) is an equality that involves the statement (13.36). This equality follows from the symmetry of the problems (13.1) and (13.30).

Let dw be an Abelian differential of the first kind. We search for the solution of the problem (13.30) in the form

$$f^{\pm}(q) = dw(q)h^{\pm}(q).$$

We obtain the boundary problem

$$h^{+}(p) = G^{*-1}(p)h^{-}(p), \qquad (h) \geqslant -(dw) \tag{13.39}$$

for vector-functions which are multiples of the divisor $-(dw)$.

The problem (13.1) is reduced to the problem for vector-differentials

$$H^{+}(p) = G(p)H^{-}(p), \qquad (H) \geqslant (dw) \tag{13.40}$$

by the change

$$H^{\pm}(q) = F^{\pm}(q)dw(q).$$

Define the functions $K^{\pm}(q)$ determined in the domains $T^{\pm}$ by the divisors $(K^{\pm}) = (dw) \cap T^{\pm}$. Then vectors $h^{\pm}(q)$ and $H^{\pm}(q)$ are represented in the domains $T^{\pm}$ by

$$h^{\pm}(q) = \frac{1}{K^{\pm}(q)}h_0^{\pm}(q), \qquad H^{\pm}(q) = K^{\pm}(q)H_0^{\pm}(q),$$

where $h_0^{\pm}$ are holomorphic vector-functions and $H_0^{\pm}(q)$ are holomorphic vector-differentials.

Then the problems (13.39) and (13.40) are reduced to the form

$$h_0^+(p) = \frac{K^+(p)}{K^-(p)} G^{*-1}(p) h_0^-(p), \tag{13.41}$$

$$H_0^+(p) = \frac{K^-(p)}{K^+(p)} G(p) H_0^-(p). \tag{13.42}$$

Taking into account that

$$\operatorname{ind}_L \det \left[\frac{K^+(p)}{K^-(P)} G^{*-1}(p) \right] = -\operatorname{ind}_L \det G + (2g-2)n,$$

we obtain, because of (13.38), the inequality

$$\begin{aligned}\dim H_G - \dim L_G \geqslant -\operatorname{ind}_L \det G \\ + (2g-2)n - ng + n = -\operatorname{ind}_L \det G + ng - n,\end{aligned}$$

which involves the relation

$$\dim L_G - \dim H_G = \operatorname{ind}_L \det G - ng + n. \tag{13.43}$$

This is the index theorem for problem (13.1). In particular, it implies the theorem statement. The formula (13.43) has two important corollaries.

THEOREM 13.8 (RIEMANN–ROCH). *Let B be a holomorphic vector bundle over M. Then*[4]

$$\dim H^0(Q(B)) - \dim H^0(Q(-B^* + K)) = \operatorname{ch}(\det B) - n(g-1). \tag{13.44}$$

Here $\det B$ is the complex line bundle with transition functions $\det h_{ij}$, and $\operatorname{ch}(\det B)$ is the value of the characteristic class of this bundle.

The second corollary is the index theorem for the operator related to the Riemann problem. Let C_L^∞ be the space of functions of the class C^∞ on $M - L$, continuous up to the contour. The operator $\mathfrak{G} f$ on C_L^∞ is defined by the formula

$$\mathfrak{G} f = \begin{cases} \bar\partial f(q), & q \in T^\pm \\ -f^+(p) + G(p) f^-(p), & p \in L. \end{cases} \tag{13.45}$$

Let D_L^∞ be the space of differentials of the class C^∞ in $T^\pm$, continuous up to the contour. Define the bilinear form on $C_L^\infty \times D_L^\infty$,

$$(\mathfrak{G} f, \varphi) = \operatorname{Re} \int_M \bar\partial f \wedge \bar\varphi + \operatorname{Re} \int_L (-f^+ + G f^-) \bar\varphi^+. \tag{13.46}$$

Then the operator conjugate to (13.45) is

$$\mathfrak{G}^* + \bar\varphi = \begin{cases} -\bar\partial \varphi(q), & q \in T^\pm \\ -\varphi^-(p) + G^*(p) \phi^+(p), & p \in L. \end{cases} \tag{13.47}$$

Therefore, solutions of the problems (13.1) and (13.30) are zeros of the operators $\mathfrak{G}$ and $\mathfrak{G}^*$, and Eq. (13.43) is the index formula of the operators G.

---

[4]The Bundle $-B^* + K$ is defined by the cocycle $\{h_{ij}^{*-1} \frac{dz_i}{dz_i}\}$, where $\{h_{ij}\}$ are transition matrices of B.

CHAPTER 4

THE RIEMANN BOUNDARY PROBLEM ON OPEN RIEMANN SURFACES

§14 Open Riemann Surfaces

A. *Finite Surfaces*

A finite Riemann surface T is a domain of a closed Riemann surface bounded by a finite number of closed analytic curves. Sometimes one considers the domain with a boundary.

The three following theorems will be used (see, [29]).

THEOREM 14.1. *Let* $g \in \Gamma(T, A^{\alpha,1})(\alpha = 0, 1)$. *Then the equation*

$$\bar{\partial} f = g \tag{14.1}$$

is solvable.[1]

In the case $\alpha = 0$ the solution is determined by the formula

$$f(q) = \frac{1}{\pi} \int_T M(p,q) dz(p) \wedge g(p), \qquad g \in \Gamma(T, A^{0,1}). \tag{14.2}$$

[1]In order to solve the equation (14.1) on a closed surface, it is necessary and sufficient that

$$\int_M g(p) \wedge \omega_j(p) = 0, \qquad j = 1, \ldots, g, \tag{14.1'}$$

where $\omega_j (j = 1, \ldots, g)$ is a basis of the Abelian differentials of the first kind.

This is proved in the same way as (9.16).

In case $\alpha = 1$, we have

$$f(q) = \frac{1}{\pi} \int_T g(p) \wedge M(q,p) dz(q) \quad , \qquad g \in \Gamma(T, A^{1,1}). \tag{14.3}$$

Note that the kernel $M(p,q)$ is constructed for any closed surface contained in T and that the characteristic divisor is placed outside the domain T. The following analog of the Runge theorem is valid.

THEOREM 14.2. *Let T_1, T_2 be two finite surfaces, $\overline{T}_1 \subset T_2 \subset T$ and*[2] *let a function $f(p)$ be holomorphic on T_1. Then for any $\varepsilon > 0$ and for any closed set $F \subset T_1$, there exists a holomorphic function $g(p)$ on T_2 such that*

$$|f(p) - g(p)| < \varepsilon, \qquad p \in F. \tag{14.4}$$

PROOF: Let some continuous linear functional be defined on the space $C(F)$ of functions which annihilates functions holomorphic in T. Then it vanishes also on the closure of this set in $C(F)$. On other hand, if every such functional annihilates some element belonging to $C(F)$, then by virtue of the Hahn–Banach theorem, this element can be approximated by the functions holomorphic on T_2. Therefore, it is sufficient to show that any linear functional on $C(F)$ annihilating functions holomorphic on T_2 also annihilates functions holomorphic in a neighbourhood of F.

Let $d\mu(p)$ be a measure on F corresponding to the considered functional. Then the value

$$\varphi(q) = \int_F M(q,p) d\mu(p)$$

is analytic in $T \backslash F$. In $T \backslash T_2, \varphi(q) \equiv 0$, since the function $M(q,p)$ is analytic in T_2 with respect to p for $q \in T_2$. Hence $\varphi(q) \equiv 0$ in $T \backslash F$.

Let now the function $h(p)$ be holomorphic in the domain $\widetilde{T} \subset T_1, \widetilde{T} \supset F$. Then, using the Cauchy theorem, we obtain

$$\begin{aligned} \int_F h(p) d\mu(p) &= \int_F d\mu(p) \frac{1}{2\pi i} \int_{\partial \widetilde{T}} h(q) M(q,p) \wedge dz(q) \\ &= \int_{\partial \widetilde{T}} h(q) dz(q) \frac{1}{2\pi i} \int_F M(q,p) d\mu(p) = 0, \end{aligned}$$

which completes the proof.

Now we turn to a study of characteristic classes on finite surfaces (§12A).

[2] The overscribed bar denotes a closure.

LEMMA. *On a finite Riemann surface T there exists an infinite differentiable function $f(p, p_0)$ having a single zero at the given point $p_0 \in T$ such that in a neighborhood p_0*

$$f(p,p_0) = [z(p) - z(p_0)]f_1(p), \qquad f_1(p_0) \neq 0,$$

where $z(p)$ is some local coordinate.

Let $T \subset M$, where M is a closed Riemann surface and a point $p_n \in M - T$. Connect the points p_0 and p_n by a curve and fix the points $p_2, \dots, p_{n-1}$ on this curve such that every pair (p_{j-1}, p_j) belongs to some coordinate neighbourhood U_j with the coordinate circle $|z| < 1$. Choose some $r, \max\{|z(p_{j-1})|, |z(p_j)|\} < r < 1$ and construct an infinite differentiable function ψ in the annulus $r \leqslant |z| < 1$ such that

$$\psi(z) = 1, \quad |z| = r,$$
$$\psi(z) = 0, \quad |z| = 1.$$

Assume that

$$f_j(p) = \begin{cases} \ln \dfrac{z - z_{j-1}}{z - z_j}, & |z(p)| < r \\ \psi(z) \ln \frac{z - z_{j-1}}{z - z_j}, & r < |z(p)| < 1, \\ 0 & p \in M \backslash U \end{cases} \qquad j = 1, \dots, u - 1.$$

Then the function

$$f(p,p_0) = \exp \sum_j f_j(p)$$

satisfies the lemma statement.

THEOREM 14.3. *Let $h = \sum_j \alpha_k p_k$ be an arbitrary divisor on the finite surface T, and let B_h be the corresponding complex line bundle. Then $c(B_h) = 0$.*

Construct the function

$$g(p) = \prod_k [f(p_1, p_k)]^{\alpha_k}, \qquad (g) = h.$$

Let $\{h_{ij}\}$ be the 1-cocycle with values in Q^* corresponding to the divisor h, $h_{ij} = h_i/h_j$.

Then $g(p) = h_i(p)g_i(p)$, where the functions $g_i(p)$ are continuous and different from zero in the domains U_i. Branches of the logarithms of g_i are defined, and $g_i = \exp k_i(p)$. We obtain that

$$d \ln h_{ij} = k_i - k_j,$$

and hence $d \ln h_{ij}$ is a cocycle in $Z^1(Q)$, and $c(B_h) = 0$ (see (12.2)).

B. *Triviality of Cohomologies on Open Riemann Surfaces*

We turn to a study of open surfaces and are primarily interested in the surfaces of the infinite genus.

We accept the almost evident fact that every open surface M possesses an exhaustion $\cdots \subset M_n \subset M_{n+1} \subset \dots, \lim M_n = M$ (see, for example, [28]). We assume that every domain M_n has a boundary ∂M_n consisting of a finite number of connected analytic curves and that M_0 is simply connected. We denote the genus of M_n by g_n; its homology basis (modulo M_n) is denoted by $k_1, \dots, k_{2g_n}$. We call this exhaustion normal and denote it by $M_n \searrow M$.

THEOREM 14.4. *Let* $g \in \Gamma(M, A^{\alpha,1}), \alpha = 0, 1$. *Then the equation*

$$\bar{\partial} f = g \tag{14.5}$$

is solvable in the class $\Gamma(M, A^{\alpha,0})$ *of single-valued infinitely differentiable forms of the type* $(\alpha, 0)$.

Let $M_n \searrow M$ be an exhaustion of M. On the surface M_n the equation

$$\bar{\partial} f_n = g$$

is solvable because of Theorem 14.1. The function $f_{n+1} - f_n$ is analytic on M_n. On account of Theorem 14.2, this function can be approximated on M_{n-1} by a function h_{n+1} holomorphic in M_{n+1},

$$|(f_{n+1} - f_n) - h_{n+1}| < \varepsilon_{n+1}, \qquad n = 2, 3, \ldots. \tag{14.6}$$

Assume

$$\hat{f}_{n+1} = f_{n+1} - h_{n+1}, \qquad n = 2, 3 \ldots. \tag{14.7}$$

All these functions satisfy the equation (14.5) on M_{n+1}, and on account of (14.6), the sequence (14.7) converges at every point of M if we suppose that $\sum_{n=2}^{\infty} \varepsilon_n < \infty$. Assuming $f(p) = \lim_{n \to \infty} \hat{f}_n(p)$, we obtain the solution of Eq. (14.5).

This implies the famous Theorem B of Cartan–Serre.

THEOREM 14.5. *If* M *is an open Riemann surface, then*

$$H^1(M, Q) = 0. \tag{14.8}$$

Consider the Dolbeault exact sequence

$$0 \longrightarrow Q \xrightarrow{i} A^0 \xrightarrow{\bar{\partial}} A^{0,1} \longrightarrow 0. \tag{14.9}$$

A^0 is fine, and hence we have the exact cohomology sequence

$$H^0(A^0) \xrightarrow{\bar{\partial}^*} H^0(A^{0,1}) \longrightarrow H^1(Q) \longrightarrow 0. \tag{14.10}$$

By virtue of Theorem 14.4, $\bar{\partial}^*$ is an epimorphism, and therefore the third term of the sequence is zero.

Now we consider the second Cousin problem. The exact sequence

$$0 \longrightarrow Q^* \longrightarrow M^* \longrightarrow \theta \longrightarrow 0 \tag{14.11}$$

(see (10.1)) implies the exact cohomology sequence

$$H^0(M^*) \longrightarrow H^0(\theta) \xrightarrow{\delta^*} H^1(Q^*). \tag{14.12}$$

In order that the second Cousin problem be solvable for the divisor $h \in H^0(\theta)$, it is necessary and sufficient that $\delta^* h = 0$. The exact sequence (cf. (12.2))

$$H^1(Q) \xrightarrow{e^*} H^1(Q^*) \xrightarrow{\nu} H^2(J) \tag{14.13}$$

and Eq. (14.8) imply that ν is a monomorphism. Therefore, the condition $\delta^* h = 0$ is equivalent to the condition $\nu(\delta^* h) = 0$, i.e., $c(B_h) = 0$. We obtain the following theorems.

THEOREM 14.6. *In order that the second Cousin problem be solvable for the divisor $h \in H^0(\theta)$, it is necessary and sufficient that the Chern class $c(B_h)$ be zero.*

THEOREM 14.7. *The complex line bundle B over M, such that $c(B) = 0$, is trivial.*

In fact, if $\{h_{kj}\}$ is the cocycle of transition functions, then by virtue of $c(B) = 0$, the cochain $\{\frac{1}{2\pi i} \ln h_{kj}\}$ is a cocycle. Because of Theorem 14.6, this cocycle is cohomological to zero, i.e.,

$$\frac{1}{2\pi i} \ln h_{kj} = h_k - h_j,$$

where h_i is holomorphic in U_i. Then

$$h_{kj} = \frac{\exp 2\pi i h_k(p)}{\exp 2\pi i h_j(p)}$$

completes the proof.

THEOREM 14.8.

$$H^1(C) = H^0(Q^1) dH^0(Q). \tag{14.14}$$

The exact sequence

$$0 \longrightarrow C \longrightarrow Q \xrightarrow{d} Q^1 \longrightarrow 0 \tag{14.15}$$

implies the exact cohomology sequence

$$H^0(Q) \xrightarrow{d^*} H^0(Q^1) \longrightarrow H^1(C) \longrightarrow H^1(Q). \tag{14.16}$$

Our statement follows from Eq. (14.8).

Note that $H^1(C)$ has a structure

$$H^1(C) = H^0(C^1)/dH^0(A^0)$$

that provides the existence of the Abelian differential with prescribed periods.

Returning to the case of the finite Riemann surface, we obtain the following result.

THEOREM 14.9. *Any complex line bundle B over a finite Riemann surface is trivial, and for any divisor h, the second Cousin problem is solvable.*

This statement follows from Theorem 14.3 and Theorem 14.6–14.8. Here we take into account that every line bundle B corresponds to some divisor h (Theorem 11.2).

Note. On any open surface M the analog of Theorem 11.9 is valid, and hence Theorem 14.9 is valid for any open surface [12].

C. *The Riemann Bilinear Relations*

Let $M_n \searrow M$ be some normal exhaustion of the surface M. Consider the family of curves $L = \{L_n\}$, $L_n = \partial M_n$ $n = 1, \ldots$. The extremal length of this family (see, for example, [28]) is a value

$$\lambda(L) = \sup \inf_{L_n \in L} \frac{\int_{L_n} |\omega|}{\int_M |\omega|^2}, \tag{14.17}$$

where the supremum is taken over the set of all differentials with a finite Dirichlet integral

$$D_M[\omega] = \int_M |\omega|^2 = \frac{1}{2i} \int_M \bar{\omega} \wedge \omega. \tag{14.18}$$

If the surface M has such exhaustion for which $\lambda(L) = 0$, the surface is said to have a zero boundary. Such surfaces are studied in many books (see, for example, [1], [28]). In particular, such surfaces are characterized by the lack of a Green's function and having zero harmonic measure of the boundary. This class of surfaces is denoted by 0_G.

Consider a surface $M \in 0_G$ with a normal exhaustion $M_n \searrow M$,

$$\partial M_n = L_n = \sum_{j=1}^{m_n} L_{nj}, \qquad \lambda(\{L_n\}) = 0.$$

We suppose that any connected component L_{nj} of L_n separates the surface M into the domains $M_{nj}^+ \subset M \backslash M_n$ and $M_{nj}^- \supset M_n$. In this case, the exhaustion is said to be an exhaustion with standard boundaries. The class of surfaces possessing an exhaustion with standard boundaries is denoted by $\tilde{0}_G$. Below we study only surfaces of the class $\tilde{0}_G$.[3]

LEMMA. *Let ω be a closed differential with a finite Dirichlet integral. Then for any connected component $L_{nj} \subset L_n$,*

$$\int_{L_{nj}} \omega = 0, \qquad j = 1, \ldots, m_n. \tag{14.19}$$

Indeed, for any $\varepsilon > 0$ there exists such $m > n$ that

$$\int_{L_m} |\omega| < \varepsilon. \tag{14.20}$$

[3] Another approach was presented by Shiba [a] and Yoshida [a] considering arbitrary open surfaces and selecting special function classes with given boundary properties.

Let $\tilde{L}_m = M_{nj} \cap L_m$. Then

$$\int_{L_{nj}} \omega = \int_{\tilde{L}_m} \omega.$$

We have

$$\left|\int_{L_{nj}} \omega\right| = \left|\int_{\tilde{L}_m} \omega\right| \leqslant \int_{\tilde{L}_m} |\omega| \leqslant \int_{L_m} |\omega| < \varepsilon,$$

Q.E.D.

THEOREM 14.10. *Let ω, φ be closed differentials with the finite Dirichlet integrals over a surface $M \subset \tilde{0}_G$. Then for any normal exhaustion $M_n \searrow M$ with standard boundaries, there exists a subsequence $n_k \longrightarrow \infty$ such that*

$$\int_M \omega \wedge \varphi = \lim_{n_k \longrightarrow \infty} \sum_{\mu=1}^{g_{n_k}} \left\{ \int_{k_{2\mu-1}} \omega \int_{k_{2\mu}} \varphi - \int_{k_{2\mu}} \omega \int_{k_{2\mu-1}} \varphi \right\}, \tag{14.21}$$

and both sides of (14.21) are finite. In particular, for $\varphi = \bar{\omega}$, we have

$$\frac{1}{2i} \int_M \bar{\omega} \wedge \omega = \lim_{n_k \longrightarrow \infty} \sum_{\mu=1}^{g_{n_k}} \operatorname{Im} \int_{k_{2\mu-1}} \bar{\omega} \int_{k_{2\mu}} \omega. \tag{14.22}$$

PROOF: Choose a point $p_j \in L_{nj}$ on any curve L_{nj}. Cut the domain M_n along the cycles $k_j (j = 1, \ldots, 2g)$ of a canonical homology basis. Then

$$\begin{gathered} \int_{M_n} \omega \wedge \varphi = \int_{M_n} \omega \wedge \Phi = \int_{L_n} \omega \Phi(p) - L_n(\omega, \varphi), \\ L_n(\omega, \varphi) = \sum_{\mu=1}^{g_n} \left(\int_{k_{2\mu-1}} \omega \int_{k_{2\mu}} \varphi - \int_{k_{2\mu}} \omega \int_{k_{2\mu-1}} \varphi \right). \end{gathered} \tag{14.23}$$

By (14.19) we have the estimate

$$\begin{aligned} \left|\int_{L_n} \omega \Phi\right| &= \left|\sum_{j=1}^{m} \left\{ \int_{L_{nj}} \omega(p)[\Phi(p) - \Phi(p_j)] + \Phi(p_j) \int_{L_{nj}} \omega \right\}\right| \\ &\leqslant \sum_{j=1}^{m} \int_{L_{nj}} |\omega| \int_{L_{nj}} |\varphi|. \end{aligned}$$

Since $\lambda(L) = 0$,

$$\int_{L_n} |\omega| \longrightarrow 0, \qquad \int_{L_n} |\varphi| \longrightarrow 0 \quad \text{as} \quad n \longrightarrow \infty. \tag{14.24}$$

Further,

$$\frac{1}{4}\left|\int_M \omega \wedge \varphi\right|^2 < \left(\frac{1}{2i}\int_M \bar{\omega} \wedge \omega\right)\left(\frac{1}{2i}\int_M \bar{\varphi} \wedge \varphi\right). \tag{14.25}$$

From this follows (14.21).

Consider the Hilbert space Γ_c of closed differentials with finite Dirichlet integrals over M with the scalar product

$$(\omega, \varphi) = \int_M \omega \wedge *\varphi, \tag{14.26}$$

where (see (2.2))

$$*(a dz + b d\bar{z}) = \frac{1}{2i}(\bar{b} dz - \bar{a} d\bar{z}).$$

The subspace of Γ_c formed by the Abelian differentials of the first kind is denoted by $H_D^0(Q^1)$.

THEOREM 14.11. *Let $T[\omega]$ be a linear functional on the space $H_D^0(Q^1)$. Then there exists a real harmonic differential η with a finite Dirichlet integral, such that*

$$\begin{aligned} T[\omega] &= \lim_{n_k \to \infty} -i \sum_{\mu=1}^{g_n}\left(c_{2\mu-1}\int_{k_{2\mu}} \omega - c_{2\mu}\int_{k_{2\mu-1}} \omega\right), \\ c_j &= \int_{kj} \eta, \qquad j = 1, 2, \ldots. \end{aligned} \tag{14.27}$$

Indeed, there exists an Abelian differential of the first kind $\theta \in H_D^0(Q^1)$, such that $T[\omega] = (\omega, \theta)$. Since $*\bar{\theta}$ is a differential of the type $(1, 0)$, we have that $(\omega, \bar{\theta}) = 0$. Therefore,

$$T[\omega] = (\omega, \theta - \bar{\theta}).$$

By assuming $\eta = \frac{1}{2}(\theta + \bar{\theta})$ and by using(14.19), we obtain Eq. (14.27).

D. *The Hodge–Royden Theorem*

Let Γ_{e0} be the linear space of exact differentials of the class C^∞ with compact support (the support depends on the differential). Denote by $\tilde{\Gamma}_{e0}$ the closure of Γ_{e0} in the topology induced by the scalar product (14.26). Let $\Gamma_h \subset \Gamma_c$ be the subspace of harmonic forms.

It is clear that $\Gamma_h \perp \tilde{\Gamma}_{e0}$. Indeed, let $\omega \in \Gamma_h$ and $\varphi \in \Gamma_{e0}$. Then $\varphi = d\Phi$, where Φ is a single-valued function. Outside of some compact set on M, this function is a constant. Let $M_n \searrow M$ be some exhaustion $\partial M_n = L_n = \sum_j L_{nj}$. We have $\varphi = 0$ on $M - M_n$ for $n > n_0$. Then for $n > n_0$,

$$\begin{aligned}(\omega, \varphi) &= \int_M \varphi \wedge *\omega = \int_{M_n} d\Phi \wedge *\omega \\ &= -\int_{M_n} \Phi \wedge d(*\omega) + \sum_j \Phi_j \int_{L_{nj}} \omega,\end{aligned}$$

where Φ_j are constant values of Φ on L_{nj}. By virtue of (14.24), the second term is zero. The first term is zero, since $d(*\omega) = 0$. Conversely, let $\omega \in \Gamma_c$. If for any $\varphi \in \Gamma_{e0}$,

$$0 = (\omega, \varphi) = \int_{M_n} d\Phi \wedge *\omega = -\int_{M_n} \Phi \wedge d(*\omega),$$

then $d(*\omega) = 0$, and ω is coclosed, and hence $\omega \in \Gamma_h$. We obtain the orthogonal decomposition in the following theorem.

THEOREM 14.12.

$$\Gamma_c = \Gamma_h \oplus \tilde{\Gamma}_{e0}. \tag{14.28}$$

This decomposition involves the existence of a harmonic differential with prescribed periods.

§15 *D*-Cohomologies

A. *D-Cohomology Groups. The Singular Group*

The triviality of cohomology groups and the existence of an analytic function determined by an arbitrary divisor on open Riemann surfaces had propelled a new approach to the study of this area. In the 1930s, L. Ahlfors and R. Nevanlinna had proposed the study of functions and differentials with a finite norm, for example, with a finite Dirichlet integral [1], [28]. On this basis, the modern theory of open Riemann surfaces was built.

In 1972 (see Rodin [o]), the author had applied this area to the study of cohomology groups and the Riemann boundary problem (see Rodin [29], [u]). We state these results in §§15–18.

Consider the covering $\mathfrak{N} = \{U_i, i \in I\}$, of the surface M by simply-connected domains. Here I is a set of indices. We suppose that there exists a number N, called a covering constant [28], such that any point $p \in M$ belongs to no more than N domains of the covering. Moreover, it is supposed that there exists a subset $I_0 \subset I$, such that $P = \{U_i, i \in I_0\}$ is a triangulation of M. The existence of such covering is proved in [28].

The D-norm of a 0-cochain $g = \{g_i\}$, with values in A^1, is a number $\|g\|_D$,

$$\|g\|_D^2 = \sum_{i \in I} \int_{U_i} |g_i|^2. \tag{15.1}$$

For a 1-cochain $\{g_{ij}\}$ with values in A^1,

$$\|g\|_D^2 = \sum_{i,j \in I} \int_{U_i \cap U_j} |g_{ij}|^2. \tag{15.2}$$

For the sheaf A^0 and its subsheaves, we assume

$$\|f\|_D = \|df\|_D, \tag{15.3}$$

and for the sheaf Q^*,

$$\|f\|_D = \|d \ln f\|_D. \tag{15.4}$$

Consider the group of cochains $Cn_D^k(F)(k = 0, 1)$ with values in the sheaf F (F is one of the sheaves A^1, A^0, Q^*, or their subsheaves). Defining the coboundary operator δ^*in the routine manner (9.3), we arrive at the following definition of the D-cohomology groups,

$$\begin{aligned} H_D^0(F) &= Z_D^0(F), \\ H_D^1(F) &= Z_D^1(F)/\delta^* Cn_D^0(F). \end{aligned} \tag{15.5}$$

It is easy to verify that an exact sequence

$$0 \longrightarrow F' \longrightarrow F \longrightarrow F'' \longrightarrow 0$$

generates the exact cohomology sequence

$$0 \longrightarrow H_D^0(F') \longrightarrow H_D^0(F) \longrightarrow H_D^0(F'') \xrightarrow{\delta^*} H^1(F') \longrightarrow \dots.$$

THEOREM 15.1. *Let* $\{g_i\} \in Cn_D^0(A^{0,1})$. *Then there exists the 0-cochain* $\{f_i\} \in Cn_D^0(A^0)$, *such that* $\bar\partial f_i = g_i$.

In the coordinate disk $|z(p)| < 1, p \in U_i, i \in I$,

$$f_i(z) = -\frac{1}{\pi} \iint_{|t|<1} g_i(t) \frac{d\sigma_t}{t - z}. \tag{15.6}$$

Indeed, we have $\bar\partial f_i = g_i$, and

$$(15.7)\qquad df_i = \left(-\frac{1}{\pi} \iint\limits_{|t|<1} g_i(t) \frac{d\sigma_t}{(t-z)^2} \right) dz + g_i d\bar z.$$

The norm of df_i over the domain U_i is equal to

$$(15.8)\qquad \|df_i\|_{L_2(U_i)} = \sqrt{2}\|g_i\|_{L_2(U_i)},$$

since the operator

$$(15.9)\qquad \Pi g_i = -\frac{1}{\pi} \iint\limits_{t<1} g_i(t) \frac{d\sigma_t}{(t-z)^2}$$

is unitary [36].

Whence it follows that the norm of the cochain $\{f_i\}$is finite, and $\{f_i\} \in Cn^0_D(A^0)$.

As follows from Theorem 15.1, the Grothendieck–Dolbeault sequence

$$(15.10)\qquad 0 \longrightarrow Q \xrightarrow{i} A^0 \xrightarrow{\bar\partial} A^{0,1} \longrightarrow 0$$

generates the exact cohomology sequence

$$(15.11)\qquad \begin{aligned} 0 \longrightarrow H^0_D(Q) \xrightarrow{i^*} H^0_D(A^0) \xrightarrow{\bar\partial^*} H^0_D(A^{0,1}) \\ \xrightarrow{\delta^*} H^1_D(Q) \\ \xrightarrow{i^1} H^1_D(A^0) \xrightarrow{\bar\partial^1} H^0_D(A^{0,1}) \longrightarrow \dots . \end{aligned}$$

THEOREM 15.2. *The group $i^1 H^1_D(Q) \subset H^1_D(A^0)$ in (15.11) is different from zero.*

Let df be a closed differential such that there exists no differential $dg \in H^0_D(C^1)$, such that $df - dg$ is exact. For example, this condition is satisfied by the differential with periods 1 along all the cycles $k_{2\mu-1}$ and periods i along all the cycles $k_{2\mu} (\mu = 1, 2, \dots)$. Because of (14.22), the Dirichlet integral of any differential with such periods is infinite. Construct the 0-cochain $\{f_i\}, df_i = df$ and the 1-cocycle $\{f_{ij}\}, f_{ij} = f_i - f_j$ with values in C. Evidently, this cocycle determines the cohomology class f nontrivial in $H^1_D(A^0)$. In the opposite case, we have $f = \delta g, g \in Cn^0_D(A^0)$, i.e.,$f_{ij} = g_i - g_j$ and $f_i - g_i = f_j - g_j$. This means that the differential $df - dg_i$ is exact, and $dg_i \in H^0_D(C^1)$. Hence the 1-cocycle $\{f_{ij}\} \in H^1_D(Q)$ has a nonzero preimage in the group $H^1_D(A^0)$.

Therefore, for D-cohomologies, the de Rham and Dolbeault theorems are invalid, since these theorems follow from the fact that $H^1(A^0) = 0$.

Let us denote

$$(15.12)\qquad \begin{aligned} S_1 &= \delta^* H^0_D(A^{0,1}) \subset H^1_D(Q), \\ S_0 &= H^1_D(Q)/\delta^* H^0_D(A^{0,1}). \end{aligned}$$

The group S_0, the nontriviality of which has been established, is called the singular group.

On account of Cartan's Theorem B (Theorem 14.5), any cocycle $\{h_{ij}\} \in Z^1_D(Q)$ is represented in the form

$$h_{ij} = H_i - H_j, \qquad \{H_i\} \in Cn^0(A^0), \qquad \bar\partial H^i = \eta \in H^0(A^{0,1}). \tag{15.13}$$

In order that the cocycle $\{h_{ij}\}$ determine a class $h \in S_1$ in the group $H^1_D(Q)$, it is necessary and sufficient that

$$H_i = F_i + f, \qquad \{F_i\} \in Cn^0_D(A^0).$$

Therefore, in order that $h \in S_1$ it is necessary that there exists a differential $\omega \in H^0_D(A^{0,1})$ such that $\eta = \omega + \bar\partial s, s \in H^0(A^0)$. We obtain the convenient sufficient condition used in the proof of Theorem 15.2.

THEOREM 15.3. *In order that the cocycle $h \in Z^1_D(Q)$ determine a nonzero element of the group S_0, it is sufficient that there exist no differential $\omega \in H^0_D(A^{0,1})$, such that*

$$\eta = \omega + \bar\partial f.$$

The integral form of this criterion is a necessary and sufficient condition describing the group S_0.

THEOREM 15.4. *In order that the cocycle $h \in Z^1_D(Q)$ determine a nonzero element of the group S_0, it is necessary and sufficient that there exist no differential $\eta \in H^0(A^{0,1})$ satisfying (15.3) and such that $H_i = F_i + f, \{F_i\} \in Cn^0_D(A^0)$.*

B. *The Serre Duality*

THEOREM 15.5. *The group S_1 defined by Eq. (15.12) is topologically dual to $H^0_D(Q^1)$ with the topology determined by the scalar product (14.26).*

In fact, the operator $\bar\partial$ generates the diagram

$$\begin{array}{ccc} H^0_D(A^0) & \xrightarrow{\ \bar\partial\ } & H^0_D(A^{0,1}) \\ {\scriptstyle duality}\downarrow & & {\scriptstyle duality}\downarrow \\ [H^0_D(A^0)]^* & \xleftarrow{\ \bar\partial^*\ } & [H^0_D(A^{0,1})]^*. \end{array} \tag{15.14}$$

The range of values of the operator $\bar\partial$ is closed. This follows from the continuity of the operation of differentiation [21]. This involves the duality of coker $\bar\partial$ and Ker $\bar\partial^*$.

Predetermining the general form of a linear functional on $H^0_D(A^{0,1})$ as

$$< \omega, \eta >= \operatorname{Re} \int_M \omega \wedge \eta, \qquad \omega \in H^0_D(A^{0,1}), \tag{15.15}$$

we obtain that η are forms of the type (1,0) with a finite Dirichlet integral and $\bar\partial^* = -\frac{\partial}{\partial \bar z} \wedge d\bar z$. This means that the spaces $H^0_D(Q^1) = \operatorname{Ker} \bar\partial^*$ and

$$\operatorname{Coker} \bar\partial^* = H^0_D(A^{0,1})/\bar\partial H^0_D(A^0) \cong S_1 \subset H^1_D(Q)$$

are dual.

THEOREM 15.6. *For any linear functional $T(dw)$ on the space $H_D^0(Q^1)$ with the scalar product (14.26), there exists the 0-cochain $\{T_i\} \in Cn_D^0(A^0)$, such that*

$$T[dw] = \int_M \bar{\partial} T_i \wedge dw. \tag{15.16}$$

By virtue of Theorem 15.5 there exists the cocycle $\{h_{ij}\} \in Z_D^1(Q)$ represented in the form

$$h_{ij} = T_i - T_j, \qquad \{T_i\} \in Cn_D^0(A^0). \tag{15.17}$$

The functional (15.16) constructed for the cocycle depends only on the cohomology class of the cocycle $\{h_{ij}\}$ that implies the theorem statement. In fact, if the cocycle $\{h'_{ij}\}$ is cohomological to $\{h_{ij}\}$ and $h'_{ij} = T'_i - T'_j$, then $T_i - T'_i = h_i + f$, $f \in Z_D^0(A^0)$, $h_i \in Cn_D^0(Q)$, and hence $\bar{\partial}(T_i - T'_i) = \bar{\partial} f$, $\quad f \in H_D^0(A^0)$. Then

$$T[dw] - T'[dw] = \int_M \bar{\partial} f \wedge dw = \lim_{n_k \to \infty} \int_{\partial M_{n_k}} f dw = 0.$$

The following theorem follows from the continuity of the operator (15.14).

THEOREM 15.7. *In order that $\omega \in \bar{\partial} H_D^0(A^0)$ it is necessary and sufficient that*

$$\int_M \omega \wedge *\varphi = 0, \qquad *\varphi \in H_D^0(Q^1), \qquad \omega \in H^0(A^{0,1}). \tag{15.18}$$

§16 D-Divisors. The Second Cousin Problem

A. *Divisor Degree*

A D-divisor h is determined by a 0-cochain with values in M^*, such that the 1-cocycle $\bar{h} = \{h_{ij}\}, h_{ij} = h_i/h_j$ in $U_i \cap U_j$ belongs to $Z_D^1(Q^*)$. The divisor h is called finite if the cohomology class $\bar{h} \in H_D^1(Q^*)$ determined by cocycle $\bar{h}$ contains a finite cocycle.

Consider the cocycle belonging to the group $Z^2(J)$,

$$\delta^* \bar{h} = \{\frac{1}{2\pi i} \ln h_{ij} - \frac{1}{2\pi i} \ln h_{ik} + \frac{1}{2\pi i} \ln h_{jk}\}. \tag{16.1}$$

This cocycle corresponds to the cocycle

$$\tilde{h} = \{\frac{1}{2\pi i} d \ln h_{ij}\} \in Z_D^1(Q^1), \tag{16.2}$$

determining the cohomology class $\tilde{h} \in H_D^1(Q^1)$.

Define the group $H^0_D(A^{1,1})$ as

$$H^0_D(A^{1,1}) = Z^0(A^{1,1}) \cap \bar\partial Cn^0_D(A^{1,0}). \tag{16.3}$$

The exact sequence

$$0 \longrightarrow Q^1 \xrightarrow{i} A^{1,0} \xrightarrow{\bar\partial} A^{1,1} \longrightarrow 0$$

generates the exact sequence

$$0 \longrightarrow H^0_D(Q^1) \longrightarrow H^0_D(A^{1,0}) \longrightarrow H^0_D(A^{1,1}) \longrightarrow H^1_D(Q^1).$$

The element $\tilde h \in H^1_D(Q^1)$ has the preimage in the group $H^0_D(A^{1,1})$ determined by the formula

$$\tau = \bar\partial \sum_{j,k\in I} \alpha_k(p) \frac{1}{2\pi i} d \ln h_{jk}(p), \quad p \in U_j, \tag{16.4}$$

where $\sum_k \alpha_k \equiv 1$ is the partition of unity subordinated to the considered covering. The form τ belongs to the group $H^0_D(A^{1,1})$ since the 0-cochain $\{T_j\}$,

$$T_j = \frac{1}{2\pi i} \sum_k \alpha_k d \ln h_{jk}, \tag{16.5}$$

belongs to the group $Cn^0_D(A^{1,0})$. Indeed, its norm is evaluated as

$$\begin{aligned}
\|\{T_j\}\|^2_D &= \sum_{j\in I} \int_{U_j} |T_j|^2 \\
&= \sum_{j\in I} \int_{U_j} | \sum_{U_k \cap U_j} \alpha_k \frac{1}{2\pi i} d \ln h_{jk}|^2 \\
&\leqslant \sum_{j\in I} \sum_{U_j\cap U_k} N \int_{U_j\cap U_k} |\frac{1}{2\pi i} d \ln h_{jk}|^2 \\
&= N \|\{h_{jk}\}\|^2_D
\end{aligned}$$

(see (15.1), (15.2)). It is clear that the form (16.4) is determined in the cohomology class $\delta^{*-1}\tilde h$ modulo an addendum of the form $\bar\partial\omega, \omega \in H^0_D(A^{1,0})$.

Analogous to §12A, we introduce the following definition. A divisor h is called a divisor of a finite degree if the integral (see §15A)

$$\deg h = \int_M \tau = \sum_{j\in I_0} \int_{U_j} \tau \tag{16.6}$$

is finite. Here the integral over M is defined as the sum of integrals over the triangles. For divisors of a finite degree, the value $\deg h$ is called the divisor degree.

THEOREM 16.1. *The definition (16.6) is correct. The degree of a finite divisor determined by the symbol $h = \sum \alpha_k p_k$ is $\deg h = \sum \alpha_k$.*

First, note that if the divisor h determines the zero element of the group $H^1_D(Q^*)$, then $\deg h = 0$. In fact, in this case the form (16.4) has the form $\tau = \bar{\partial}\omega, \omega \in H^0_D(A^{1,0})$. We obtain

$$\int_M \tau = \lim_{n_k \to \infty} \int_{\partial M_{n_k}} \omega = 0,$$

since the extremal length of the family $L = \{\partial M_n\}$ is equal to zero.

We calculate the degree of a finite divisor.

$$\begin{aligned}\int_M \tau &= \sum_{j \in I_0} \int_{U_j} \tau = \sum_{j \in I_0} \int_{\partial U_j} T_j \\ &= \sum_{j \in I_0} \int_{\partial U_j} \sum_{U_j \cap U_k} \alpha_k \frac{1}{2\pi i} d \ln h_{jk} \\ &= \sum_{j \in I_0} \frac{1}{2\pi i} \int_{\partial U_j} d \ln h_j \\ &\quad - \sum_{j \in I_0} \sum_{U_j \cap U_k} \frac{1}{2\pi i} \int_{(\partial U_j) \cap U_k} \alpha_k d \ln h_k .\end{aligned}$$

All sums here are finite. The first one is equal to $\deg h = \sum \alpha_k$, and the second one is equal to zero, since every portion of the boundary ∂U_j of U_j is passed twice in the opposite directions. If the boundaries of U_i and U_k coincide, $\alpha_k = 0$ on the corresponding part of the boundary.

B. *Infinite Divisors*

The special covering. Let $M_n \searrow M$ be a normal exhaustion of the surface $M, M_n = L_n$. Denote the annular domain $M_{n+1} - M_n - L_n$ by $V_{n+1}, n > 0; V_0 = M_0$. Further, let Δ_n be an annular domain, $\Delta_n \subset M_n$, bounded by the contours L_n and L'_n, where L'_n is a curve homotopic to L_n, $L'_n \subset M_n, L_n \cap L'_n = 0$. Assume $W_{n+1} = V_{n+1} + L_n + \Delta_n, W_0 = M_0$. The domains $V_n, W_n (n = 0, 1, \ldots)$ are called zones. Choose two sets of indices, I_n and J_n. We consider the coverings $\{U_i, i \in I_n\}, U_i \subset V_n$, and $\{U_a, a \in J_n\}, U_a \subset W_n$, of domains V_n and W_n, respectively, by simply-connected domains $\{U_i\}$ and $\{U_a\}$. The union of all domains $\{U_i, i \in I_n\}, \{U_a, a \in J_n\} n = 0, 1, 2, \ldots$ forms the special covering of the surface M.

We consider 0-cochains corresponding to the special covering and with the form

$$h_i^n = h_n(q), \qquad q \in V_n, \qquad i \in I_n,$$
$$h_a^n = h_n(q), \qquad q \in W_n, \qquad a \in J_n. \tag{16.7}$$

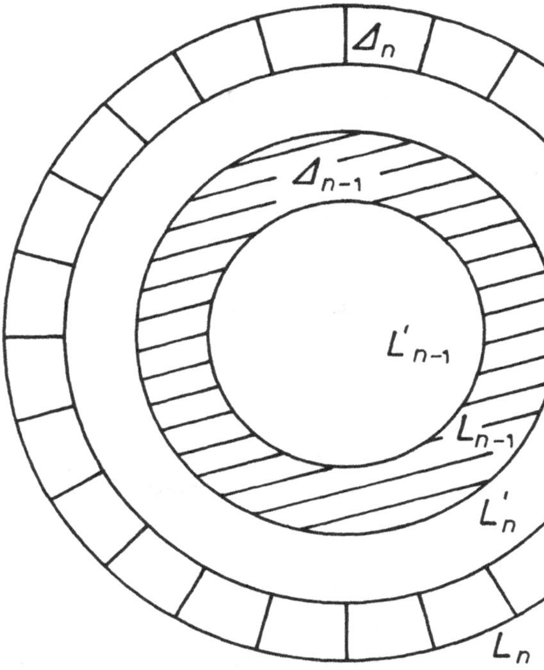

Figure 10
The domain W_n

The example of an infinite D-divisor of the infinite degree. The divisor h is defined as follows. Let the point $p_{2n} \in V_{2n}, p_{2n} \in \Delta_{2n}$. Assume

$$
\begin{aligned}
&h_{2n+1}(q) \equiv 1, \qquad && q \in W_{2n+1},\\
&h_{2n}(q) = p_{2n}(q), \qquad && q \in W_{2n},
\end{aligned}
\tag{16.8}
$$

where the function $p_{2n}(q)$ is holomorphic in the domain W_{2n} and has a zero of the first order at the point p_{2n}.

The transition functions corresponding to the divisor h are

$$
\begin{aligned}
&h_{ij}^{2n,2n+1} = h_{ia}^{2n,2n+1} = p_{2n}(q), \qquad q \in U_i^{2n} \cap U_j^{2n+1},\\
&\qquad q \in U_i^{2n} \cap U_a^{2n+1}, \qquad i \in I_{2n}, \qquad j \in I_{2n+1}, \qquad a \in J_{2n+1};\\
&h_{ij}^{2n+1,2n+2} = h_{ia}^{2n+1,2n+2} = (p_{2n+2}(q))^{-1}, \qquad q \in U_i^{2n+1} \cap U_j^{2n+2},\\
&\qquad q \in U_i^{2n+1} \cap U_a^{2n+2}, \qquad i \in I_{2n+1}, \qquad j \in I_{2n+2}, \qquad a \in J_{2n+2},\\
&h_{ia}^{k,k} \equiv 1, \qquad q \in U_i^k \cap U_a^k, \qquad i \in I_k, \qquad a \in J_k.
\end{aligned}
$$

Below we omit the indices i, j, a and use numbers of zones only. These indices we write as

$$h_{2n,2n+1}(q) = p_{2n}(q),$$
$$h_{2n+1,2n+2}(q) = (p_{2n+2}(q))^{-1},$$
$$h_{kk}(q) = 1.$$

The norm of the cochain $\{h_{ls}\}$ (l, s are zone numbers) is

$$\|\{h_{ls}\}\|_D^2 = \sum_{k=1}^{\infty} \int_{\Delta_k} |d \ln p_k(q)|. \tag{16.9}$$

Choose the domains Δ_n so narrow that the series (16.9) converges. On the other hand,

$$\deg h = \lim_{n_k \to \infty} \int_{M_{n_k}} \tau = \lim_{n_k \to \infty} n_k = \infty.$$

The example of an infinite D-divisor of the finite degree. Let $f(q)$ be a holomorphic function on M with an infinite Dirichlet integral. Using the special covering, we assume

$$\begin{aligned} h_{2n+1}(q) &= 1, & q \in W_{2n+1}, \\ h_{2n}(q) &= \exp f(q), & q \in W_{2n}. \end{aligned} \tag{16.10}$$

The domains Δ_n are chosen sufficiently narrow so that the norm of the cocycle $\bar{h}$ (see (16.9)) be finite. We shall calculate the divisor h degree. Let $\sum \beta_s \equiv 1$ be a partition of unity subordinated to the special covering. Assume that

$$\alpha'_n(q) = \sum_{j \in I_n} \beta_j(q), \qquad \alpha_n(q) = \sum_{a \in J_n} \beta_a(q). \tag{16.11}$$

The functions α_n have the following properties

$$\alpha_{n+1} \equiv 1, \qquad q \in L_n; \qquad \alpha_{n+1} \equiv 0, \qquad q \in L'_n, \tag{16.11'}$$

since the contours L_n are covered by the domains $U_a, a \in J_{n+1}$, and the contours L_n are covered by the domains $U_a, a \in J_n, U_j, j \in I_n$.

The form (16.4) is determined by the relation

$$\tau = \begin{cases} \frac{1}{2\pi i} \bar{\partial}[\alpha_{n+1} \partial \ln h_{n,n+1}] & q \in \Delta_n \\ 0 & q \in V_n \backslash \Delta_n, \end{cases} \tag{16.12}$$

since the transition functions are equal to unity in the domains $V_n \setminus \Delta_n$. Hence

$$\begin{aligned} \deg h &= \frac{1}{2\pi i} \sum_{n=0}^{\infty} \int_{\Delta_n} \bar{\partial}[\alpha_{n+1}(q) \partial \ln h_{n,n+1}(q)] \\ &= \frac{1}{2\pi i} \sum_{n=0}^{\infty} \{ \int_{L_n} \alpha_{n+1}(q) \bar{\partial} \ln h_{n,n+1}(q) \\ &\quad - \int_{L'_n} \alpha_{n+1}(q) \bar{\partial} \ln h_{n,n+1}(q) \}. \end{aligned}$$

Since $\alpha_{n+1} \equiv 1$ on L_n and $\alpha_{n+1} \equiv 0$ on L'_n, we obtain

(16.13) $$\deg h = \sum_{n=0}^{\infty} (-1)^n \int_{L_n} df = 0.$$

We shall show that our divisor is infinite, i.e., is the cohomological nowhere finite cochain $\{g_n\}$. Indeed, in this case,

$$\begin{aligned} \exp f &= g_{2n}(q)s(q) \quad q \in V_n, \\ 1 &= g_{2n+1}(q)s(q)\, q \in V_{n+1}, \end{aligned} \quad s \in H^0_D(Q^*)$$

This relation means that the Dirichlet integral of the function f is finite over the union of the domains V_{2n}. This situation can be prevented by proper choice of function $f(q)$.

C. *S-Divisors*

We shall study a class of infinite divisors playing the major role in Abel's theorem.

We consider a finite divisor $h_n = \sum_k \alpha_{nk} p_{nk}, \deg h_n = 0, h_n \cap \Delta_n = 0$ in every zone $V_n (n = 0, 1 \ldots)$. Let $Q_n(q)$ be an analytic function defined on the surface M_n, having logarithm singularities with coefficients α_{nk} at the points p_{nk} and single-valued on the surface M_n cutting along paths connecting the points of the divisor h_n. Determine the cochain $\{h_i\}$ by values in the domains V_n,

(16.14) $$h_n(q) = \exp 2\pi i Q_n(q), \qquad q \in V_n.$$

The finiteness of the norm of the cocycle $\bar{h}$ (see (16.6)) can be provided by a proper choice of the domains Δ_n.

THEOREM 16.2. *In order that the 1-cocycle $\tilde{h}$ determined by relations (16.2) and (16.14) determine the cohomology class belonging to the group S_1, it is sufficient that the series*

(16.15) $$\sum_{n=0}^{\infty} \left(\sum_k \alpha_{nk} w(p_{nk}) \right) < \infty$$

converge for any Abelian differential of the first kind $dw \in H^0_D(Q^1)$, and the functional (16.15) be bounded on $H^0_D(Q^1)$.

In order that the cocycle $\tilde{h}$ determines the element of the group S_1, it is necessary and sufficient, by Theorems 15.5 and 15.6, that this cocycle be represented in the form

(16.16) $$\frac{1}{2\pi i} \ln h_{jk} = T_j - T_k, \qquad \{T_j\} \in Cn^0_D(A^0),$$

where

(16.17) $$\int_M \bar{\partial} T_j \wedge dw < \infty, \qquad dw \in H^0_D(Q^1).$$

The cochain $\{T_i\}$ may be constructed by the formula

$$T_i = Q_i - \sum_k \tilde{\alpha}_k Q_k, \qquad \tilde{\alpha}_k = \alpha_k + \alpha'_k. \tag{16.18}$$

We shall verify that $\{T_i\} \in Cn^0_D(A^0)$, if Eq. (16.15) is valid. Indeed, from (16.11)

$$\begin{aligned} T_n &= 0, && q \in V_n \backslash \Delta_n, \\ T_n &= \alpha_{n+1}(Q_n - Q_{n+1}) && q \in \Delta_n. \end{aligned} \tag{16.19}$$

Now we calculate the value of functional (16.17). Let

$$s(q) = \sum_{k=0}^{\infty} \tilde{\alpha}_k(q) Q_k(q). \tag{16.18$'$}$$

Then

$$\begin{aligned} \int_M \bar{\partial} T \wedge dw &= \sum_{n=0}^{\infty} \int_{V_n} \bar{\partial} T_n \wedge dw \\ &= \sum_{n=0}^{\infty} \int_{\partial V_n} T_n \wedge dw \\ &= \sum_{n=0}^{\infty} \left\{ \int_{\partial V_n} Q_n dw - \int_{\partial V_n} s dw \right\}. \end{aligned}$$

The first integral is easily calculated and is equal to

$$\int_{\partial V_n} Q_n dw = \sum_k \alpha_{nk} w(p_{nk}), \qquad n = 0, 1, \ldots. \tag{16.20}$$

From (16.11′), the second integral is equal to

$$\begin{aligned} \sum_{m=1}^{n} \int_{\partial V_m} s dw &= \int_{\partial M_n} s dw \\ &= \int_{L_n} Q_{n+1} dw = 0, \end{aligned} \tag{16.21}$$

since $Q_{n+1}(q)$ is a function holomorphic in the domain M_n. Therefore

$$T[dw] = \sum_{n=0}^{\infty} \left(\sum_k \alpha_{nk} p_{nk} \right) < \infty. \tag{16.22}$$

The divisor h determined by the cochain $\{h_i\}$ is termed special (S-divisor) if one is defined for the special covering and has the form

$$h_n(q) = \exp 2\pi i\{Q_n(q) + f_n(q)\}, \tag{16.23}$$

where $f_n \in Cn^0(Q), \delta^*\{f_n\} \in S_1$, the 0-cochain $\exp 2\pi i Q_n$ (16.14) satisfies the condition (16.15), and the functional (16.15) is bounded on $H^0_D(Q^1)$.

D. *The Second Cousin Problem*

We formulate the second Cousin problem in the following manner. Let $\{h_i\}$ determine the D-divisor. It is necessary to search for a D-cochain $\{f_i\}$ with values in Q^*, such that

$$h_{ij} = \frac{h_i}{h_j} = \frac{f_i}{f_j}. \tag{16.24}$$

The exact sequence

$$0 \longrightarrow J \xrightarrow{i} Q \xrightarrow{e} Q^* \longrightarrow 0, \qquad e(f) = \exp 2\pi i f(p) \tag{16.25}$$

generates the exact cohomology sequence

$$\ldots \longrightarrow H^1(J) \xrightarrow{i^*} H^1_D(Q) \xrightarrow{e^*} H^1_D(Q^*) \xrightarrow{\nu} H^2(J). \tag{16.26}$$

If the cocycle $\bar{h} = \{h_{ij}\}$ determines the zero cohomology class $\bar{h}$ in the group $H^1_D(Q^*)$, then $\nu(\bar{h}) = 0$. If the latter condition is valid, there exists a cohomology class $e^{*-1}(\bar{h}) \in H^1_D(Q)$ determined by the cocycle $\{1/2\pi i \ln h_{ij}\}$.

In order that the second Cousin problem be solvable, it is necessary and sufficient that

$$e^{*-1}(\bar{h}) \in i^* H^1(J). \tag{16.26'}$$

We shall consider at greater length the case when h is an S-divisor. The condition $\nu(\bar{h}) = 0$ is valid. Indeed, construct the form (16.14). It is clear that only the terms

$$\ln h^{n,n+1}_{ia}, \qquad h^{n,n+1}_{ia} = h^n_i / h^{n+1}_a$$

are different from zero in the domains $\Delta_n, n = 0, 1, \ldots$. Taking into account (16.11), we find that

$$\tau = \begin{cases} \frac{1}{2\pi i}\bar{\partial}\alpha_{n+1}\partial \ln h_{n,n+1}, & q \in \Delta_n \\ 0, & q \in V_n \backslash \Delta_n. \end{cases}$$

Hence,

$$\begin{aligned} \deg(h) &= \sum_{n=0}^{\infty} \int_{U_n} \tau = \sum_{n=0}^{\infty} \frac{1}{2\pi i} \int_{\Delta_n} \bar{\partial}\alpha_{n+1}\partial \ln \frac{h_n}{h_{n+1}} \\ &= \sum_{n=0}^{\infty} \left\{ \frac{1}{2\pi i} \int_{L_n} \alpha_{n+1} d \ln \frac{h_n}{h_{n+1}} - \frac{1}{2\pi i} \int_{L'_n} \alpha_{n+1} d \ln \frac{h_n}{h_{n+1}} \right\} \\ &= \sum_{n=0}^{\infty} \frac{1}{2\pi i} \int_{L_n} d \ln \frac{h_n}{h_{n+1}}, \end{aligned}$$

since $\alpha_{n+1}(q) = 0$ as $q \in L'_n$ (see (16.11′)). Further

$$\frac{1}{2\pi i}\int_{L_0} d\ln h_0 = \frac{1}{2\pi i}\int_{L_n - L_{n-1}} d\ln h_n = 0, \qquad n > 0,$$

since the sum of residues of h_n in V_n is equal to zero. Therefore,

$$\deg(h) = 0.$$

The divisor h is determined by the cochain $\{h_n\}$,

$$h_n = \exp 2\pi i\{Q_n + f_n\}, \tag{16.27}$$

where $\delta^*\{f_n\} \in S_1$. Hence, by (15.12)

$$\begin{aligned} f_n(q) &= m_n(q) + t(q), & \{m_n\} &\in Cn^0_D(A^0), \\ Q_n(q) &= T_n(q) + s(q), & t, s &\in Z^0(A^0), \end{aligned} \tag{16.28}$$

where $\{T_n\}$ is the cochain (16.18), and $s(q)$ is determined by (16.18′).

Consider the functionals

$$\begin{aligned} T[dw] &= \int_M \bar\partial T_n \wedge dw, \\ m[dw] &= \int_M \bar\partial m_n \wedge dw, \\ S[dw] &= T[dw] + m[dw], \qquad dw \in H^0_D(Q^1). \end{aligned} \tag{16.29}$$

By virtue of Theorem 14.11 there exists a real harmonic differential η_S, such that $S[dw] = 4(dw, *\eta_S)$ (see (14.26)).

THEOREM 16.3. *Let h be an S-divisor. In order that the second Cousin problem be solvable it is necessary and sufficient that all periods of the differential η_S be integers.*

Necessity. Let

$$\{\frac{1}{2\pi i}\ln h_{jk}\} \in i^* H^1(J) \subset H^1_D(Q).$$

Then

$$\frac{1}{2\pi i}\ln h_{jk} = \alpha_j - \alpha_k + m_{jk}, \qquad \{\alpha_j\} \in Cn^0_D(Q),$$

where m_{jk} are integers. Since h is an S-divisor, we have from (16.23) and (16.16)

$$\frac{1}{2\pi i}\ln h_{jk} = S_j - S_k, \qquad \{S_j\} \in Cn^0_D(A^0).$$

Then $\exp 2\pi i(S_j - \alpha_j) = f(q)$ is a single-valued function on M, and hence

$$d(S_j - \alpha_j) = \frac{1}{2\pi i} d\ln f(q) \in Z_D^0(A^1).$$

Then $\bar{\partial} S_j = \frac{1}{2\pi i}\bar{\partial}\ln f(q)$ and

$$S[dw] = \int_M \bar{\partial} S_j \wedge dw = \int_M \frac{1}{2\pi i} d\ln f \wedge dw. \tag{16.30}$$

The differential η_S is defined by the functional

$$S[dw] = -4\int_M dw \wedge * * \eta_S = \int_M dw \wedge \eta_S. \tag{16.31}$$

By (14.28′) we obtain the relation

$$\eta_S = -\frac{1}{2\pi i} d\ln f + dr,$$

where r is a single-valued function, and hence the periods of η_S are integers.

Sufficiency. If the periods of η_S are integers, then $\eta_S = \frac{1}{2\pi i} d\ln f \in H_D^0(C^1)$, where f is a single-valued function. We have

$$\int_M [\bar{\partial}(T_n + m_n) - \frac{1}{2\pi i} d\ln f) \wedge dw = 0,$$

$$dw \in H_D^0(Q^1).$$

Because of Theorem 15.7, this means that

$$\bar{\partial}(T_n + m_n) - \frac{1}{2\pi i} d\ln f = \alpha dz + \bar{\partial} g,$$

where αdz is an arbitrary form of the type $(1,0)$, and $g \in H_D^0(A^0)$. Hence

$$T_n + m_n = \frac{1}{2\pi i}\ln f + g + \beta_n,$$

where $\{\beta_n\} \in Cn_D^0(A^0)$, since the left-hand side has a finite Dirichlet integral. Then (see (16.28)),

$$\frac{1}{2\pi i}\ln h_{kj} = (T_k + m_k) - (T_j + m_j) = m_{kj} + (\beta_k - \beta_j)$$

where m_{ij} are integers determined by the branches of $\ln f(q)$ in the domains V_i, V_j that implies the statement of the theorem.

We obtain the following theorem

THEOREM 16.4 (ABEL). *In order that the second Cousin problem be solvable for an S-divisor h it is necessary and sufficient that there exist such integers c_j that*

$$\sum_{n=0}^{\infty}\left(\sum_{k}\alpha_{nk}w(p_{nk})\right)$$
$$= \lim_{n_k\to\infty}\sum_{\mu=1}^{g_{n_k}}\left(c_{2\mu-1}\int_{k_{2\mu}} dw - c_{2\mu}\int_{k_{2\mu-1}} dw\right) - m[dw],$$
$$dw \in H_D^0(Q^1).$$

Indeed,

$$S[dw] = (dw, *\eta'_S), \qquad \eta'_S = 4\eta_S.$$

Taking into account (16.22) and Theorem 14.11, we obtain

$$T[dw] = \sum_{n=0}^{\infty}(\sum_{k}\alpha_{nk}w(p_{nk})),$$

and

$$(dw, *\eta'_S) = \lim_{n_k\to\infty}\sum_{\mu=1}^{g_{n_k}}(c_{2\mu-1}\int_{k_{2\mu}} dw - c_{2\mu}\int_{k_{2\mu-1}} dw),$$

where the integers c_j are periods of η'_S that involve (16.32).

§17 The Riemann Problem. Solvability

A. *The Problem Statement. The Bundle B_G*

We suppose that M has an exhaustion with standard boundaries as above. The contour L is assumed to be compact and closed and to separate the surface into the domains $T^{\pm}$, that are, in general, noncompact (see Notes in §5), $G(p) \neq 0$ on L.

We shall study the problem

$$F^{+}(p) = G(p)F^{-}(p). \tag{17.1}$$

In the simplest variant, we search for the solution with a finite Dirichlet integral

$$\int_{M\setminus\Delta(L)} |dF|^2 \subset \infty. \tag{17.2}$$

Here$\Delta(L)$ is some neighbourhood of the contour L. As long as the solutions with infinite number of zeros and poles are most interesting in the noncompact case, we expand this statement to provide a possibility to study such solutions.

Let $\{U_i\}$ be some covering of M satisfying the conditions of §15 and let $\{g_i\}$ be some fixed 0-cochain. The functions g_i are meromorphic in the domains U_i and satisfy the boundary condition

$$g_i^+(p) = G(p)g_i^-(p), \qquad p \in L \cap U_i \tag{17.3}$$

if $U_i \cap L \neq 0$. We assume that the transition functions $g_{ij} = g_j/g_i$ are holomorphic and different from zero in $U_i \cap U_j$. The complex line bundle determined by the transition functions is denoted by B_G.

A holomorphic solution of the problem (17.1) is a holomorphic cochain $\{f_i\}$, such that

$$\frac{f_i}{f_j} = g_{ij}, \tag{17.4}$$

with a finite Dirichlet integral

$$\sum_{i\in I} \int_{U_i} |df_i|^2 \subset \infty. \tag{17.5}$$

It is clear that $\Phi(q) = g_i(q)f_i(q)$ is a function analytic on M which is multiples of the divisor $g = \{g_i\}$ and satisfies the bounded condition (17.1). If the cochain $\{g_i\}$ is finite and all functions g_i are holomorphic and different from zero, we obtain the problem statement described in the beginning of this section.

· Simultaneously with problem (17.1), we consider the problem

$$\Psi^+(p) = G^{-1}(p)\Psi^-(p) \tag{17.6}$$

for analytic differentials with a finite Dirichlet integral

$$\int_{M\setminus\Delta(L)} |\Psi|^2 \subset \infty. \tag{17.7}$$

The bundle B_G suggests a broader treatment of a solution. The cochain $\{\psi_i\} \in Cn_D^0(Q^1)$ is called a holomorphic solution of the problem (17.6) if

$$\psi_i/\psi_j = g_{ij}^{-1}. \tag{17.8}$$

$\Psi = \psi_i g_i^{-1}$ is an analytic differential which is multiples of the divisor g and satisfies the boundary condition (17.6). If the functions g_i are uniformly bounded, then Ψ has a finite Dirichlet integral.

We obtain the following theorem.

THEOREM 17.1. *The space of solutions of problem (17.1) is* $H_D^0(Q(B_G))$; *the space of solutions of problem (17.6) is* $H_D^0(Q^1(-B_G))$.

B. *The Existence of a Solution*

Let h be some divisor on M determined by the cochain $\{h_i\}$, and the bundle B_h is determined by the transition functions $\{h_{ij}\}, h_{ij} = h_i/h_j$. Then the sheaf $Q_h(B)$ of germs of meromorphic sections of the line bundle B which are multiples of the divisor $-h$ is isomorphic to the sheaf $Q(B+B_h)$ of germs of meromorphic sections of $B+B_h$. We design elements of the group $H^0_D(Q(B_G+B_h))$ as solutions of the problem (17.1) which are multiples of the divisor $-h$ and elements of the group $H^0_D(Q^1(-B_G-B_h))$ as solutions of the problem (17.6) which are multiples of the divisor h.

Let $h \geqslant 0$ be a finite divisor determined by the symbol $\sum \alpha_k p_k, \alpha_k \geqslant 0$. Consider the exact sequence

$$0 \longrightarrow Q(B) \xrightarrow{i} Q(B+B_h) \xrightarrow{\pi} Q(h,B) \longrightarrow 0, \tag{17.9}$$

where the homomorphism i is defined by the formula $i\{f_j\} = \{h_j f_j\}$. We obtain the exact cohomology sequence

$$0 \longrightarrow H^0_D(Q(B)) \xrightarrow{i^*} H^0_D(Q(B+B_h)) \xrightarrow{\pi^*} H^0(h,B) \xrightarrow{\delta^*} H^1_D(Q(B)). \tag{17.10}$$

Consider now the Grothendieck–Dolbeault sequence

$$0 \longrightarrow Q(B) \longrightarrow A^0(B) \xrightarrow{\bar\partial} A^{0,1}(B) \longrightarrow 0. \tag{17.11}$$

LEMMA. *If* $\{\omega_i\} \in Cn^0_D(A^{0,1}(B))$, *then there exists such a 0-cochain* $\{\tau_i\}$ *that* $\bar\partial\tau_i = \omega_i, \{\tau_i\} \in Cn^0_D(A^0,(B))$.

In the local coordinate plane $z(p), p \in U_i$, we have

$$\tau_i(z(q)) = -\frac{1}{2\pi i}\int\limits_{z(U_i)} \omega_i(z(p)) \wedge \frac{dz(p)}{z(p)-z(q)}, \qquad q \in U_i.$$

The finiteness of the norm of the cochain $\{\tau_i\}$ is proved as in Theorem 15.1.

The exactness of the sequence

$$\ldots \longrightarrow H^0_D(A^0(B)) \xrightarrow{\bar\partial^*} H^0_D(A^{0,1}(B)) \xrightarrow{\nu^*} H^1_D(Q^1(B)) \tag{17.12}$$

follows from the lemma.

Denote

$$S_1 = \nu^* H^0_D(A^{0,1}(B)) \cong H^0_D(A^{0,1}(B))/\bar\partial^* H^0_D(A^0(B)). \tag{17.13}$$

As follows from Theorems 15.2 and 15.3, the group S_1 does not coincide with any group $H^1_D(Q(B))$. The structure of the singular group $S_0 = H^1_D(Q(B))/S_1$ for the case of trivial B was studied in §15A. In the general case, analogous theorems are valid.

THEOREM 17.2. *The image of the group $H^0(h, B)$ in (17.10) belongs to the group S_1,*

$$\delta^* H^0(h, B) \subset S_1 \subset H^1_D(Q(B)).$$

Indeed, let $f \in H^0(h, B)$. Its preimage in the group $Cn^0_D(Q(B + B_h))$ is some cochain $\{f_i\}, f_i = g_{ij}h_i f_j/h_j$, where $\{g_{ij}\}$ are transition functions of B. Then $i^{*-1}\{f_i\}$ is the cochain $\{f_i/h_i\}$ with values in $Q_{-h}(B)$. The coboundary of this cochain in the coordinate z_i is $\{f_{ij}\}$,

$$f_{ij} = f_i/h_i - g_{ij} f_j/h_j.$$

These functions are regular at the points of the divisor h, since the principal parts R_i and R_j of f_i/h_i and f_j/h_j are connected by the transform $R_i = g_{ij}R_j$. Outside some compact domain of M, all functions $h_i = 1$.

We shall show that the cochain $\{f_{ij}\} \in H^1_D(Q(B))$ is cohomological to zero in the group $H^1_D(A^0(B))$. Let$\sum \alpha_i \equiv 1$ be a partition of unity subordinated to the covering considered. Examine the cochain $\{m_i\} \in Cn^0_D(A^0(B))$,

$$m_i = \sum_{k \in I} \alpha_k f_{ik} = \frac{f_i}{h_i} - \sum_{k \in I} \alpha_k \frac{g_{ik} f_k}{h_k}.$$

Then the coboundary of the cochain $\{m_i\}$ is determined by the cocycle $\{m_{ij}\}$,

$$\begin{aligned} (17.14) \quad m_{ij} &= m_i - g_{ij} m_j \\ &= \frac{f_i}{h_i} - \sum_{k \in I} \alpha_k \frac{g_{ik} f_k}{h_k} - \frac{f_j g_{ij}}{h_j} + \sum_{k \in I} \alpha_k \frac{g_{jk} g_{ij} f_k}{h_k} = \frac{f_i}{h_i} - \frac{f_j g_{ij}}{h_j} = f_{ij}. \end{aligned}$$

This involves $\{f_{ij}\} \in \mathrm{Im}\nu^*$ in (17.12).

THEOREM 17.3. *If the problem (17.6) has no holomorphic solution, the problem (17.1) has a solution with a finite number of poles.*

Let $h \geqslant 0$ be a finite divisor. Then $\dim H^0(h, B_G) = \deg h$ (cf. with (11.10)). The sequence (17.10) means that at least one of the groups $H^0_D(Q(B_G + B_h))$ and $H^1_D(Q(B_G))$ is different from zero. If $\mathrm{Im}\delta^*$ is different from zero, then by Theorem 17.2, the group

$$S_1 = H^0_D(A^{0,1}(B_G))/\bar{\partial} H^0_D(A^0(B_G))$$

is different from zero (cf. Theorem 11.1). The group S_1 is dual to the group $H^0_D(Q^1(-B_G))$. Therefore, at least one of the groups

$$H^0_D(Q(B_G + B_h)),\ H^0_D(Q^1(-B_G))$$

is different from zero, which implies the statement of the theorem.

We obtain a convenient criterion of the solvability of the Riemann problem.

THEOREM 17.4. *If at least one of the groups $H^1_D(Q(B_G)), H^0_D(Q^1(-B_G))$ is finite-dimensional, then the problem(17.1) has a solution having a finite number of poles.*

Indeed, we shall choose a divisor h such that $\deg h > \dim H^0_D(Q^1(-B_G))$. Then $\operatorname{Ker}\delta^* \neq 0$, and hence $\pi^{*-1}H^0(h, B)$ is nontrivial in the group $H^0_D(Q(B_G + B_h))$.

Theorem 17.3 may be made more accurate. Rewrite the sequence (17.10) in the form

$$(17.15)\quad 0 \longrightarrow H^0_D(Q(B_G + B_h))/i^* H^0_D(Q(B_G)) \xrightarrow{\pi^*} H^0(h, B_G)) \xrightarrow{\delta^*} \operatorname{Ker}\{H^1_D(Q(B)) \xrightarrow{i^*} H^1_D(Q(B_G + B_h))\} \longrightarrow 0.$$

This implies the following theorem.

THEOREM 17.5. *The relation*

$$(17.16)\quad \dim H^0_D(Q(B_G + B_h))/i^* H^0_D(Q(B_G)) + \dim \operatorname{Ker}\{H^1_D(Q(B_G) \xrightarrow{i^*} H^1_D(Q(B_G + B_h))\} = \deg h$$

is valid.

This statement is equivalent to the Riemann–Roch theorem written in the form

$$(\dim H^0(Q(B_h)) - 1) + (g - \dim Q^1(-B_h)) = \deg h$$

if the genus of the surface is finite.

In the general case, we have a problem of the infinite index.

C. *The Cauchy Index. The Solvability Conditions*

In this section we study Riemann problems for which B_G is a D-bundle, i.e., the cocycle $\{g_{ij}\}$ has a finite D-norm and hence determines some cohomology class $g \in H^1_D(Q^*)$. Let $\sum \alpha_k \equiv 1$ be a partition of unity, $I_0 \subset I$ be the indices subset determining a triangulation of M (§15), and

$$(17.17)\qquad \tau_j = \sum_{k\in I} \alpha_k \frac{1}{2\pi i} d \ln g_{jk}.$$

The value

$$(17.18)\qquad \kappa = \int_M \bar\partial \tau_j = \sum_{j\in I_0} \int_{U_j} \bar\partial \tau_j$$

is called the Cauchy index of the problem (17.1). It is easy to see that this definition generalizes the concept of the divisor degree (§16). If the integral (17.7) diverges, the index is defined as infinite. If the cochain $\{g_i\}$ is finite, then

$$\kappa = \sum_{j\in I_0} \int_{U_j} \bar\partial\tau_j = \sum_{j\in I_0} \int_{\partial U_j} \tau_j = \sum_{j\in I_0} \frac{1}{2\pi i} \int_{\partial U_j} d\ln g_j + \sum_{j\in I_0} \frac{1}{2\pi i} \int_{L\cap U_j} d(\ln g_j^+ - \ln g_j^-) + \sum_{j\in I_0} \frac{1}{2\pi i} \int_{U_k\cap\partial U_j} \alpha_k d\ln g_k.$$

The first sum of the right-hand side is equal to the difference between the numbers of zeros and poles of the functions g_j denoted by $\tilde{\kappa}$. The second term is equal to

$$\kappa_0 = \frac{1}{2\pi i} \int_L d \ln G.$$

It is the index of the boundary condition. In the third term, as always, U_j are triangulation domains, and U_k are any domains of the covering. This sum is equal to zero since every portion of ∂U_j is passed twice in the opposite directions. Therefore, the Cauchy index is equal to $\kappa = \kappa_0 + \tilde{\kappa}$ in this case.

A cochain $\{f_j\}$ is different from zero on the ideal boundary of the surface M if for any normal exhaustion $M_n \searrow M$ there exists such $\varepsilon > 0$ that $|f_j| > \varepsilon$ on $M \backslash M_n, n > n_0$. A cochain $\{f_j\}$ is bounded on the ideal boundary if there exists such C that $|f_j| < C$ on $M \backslash M_n$.

THEOREM 17.6. *Let the Cauchy index of the problem (17.1) be finite and equal to κ. Then a holomorphic solution of this problem different from zero on the ideal boundary has κ zeros.*

PROOF: Let $\{f_j\}$ be a solution of (17.1). Then $g_{ij} = f_i/f_j$, and

$$\tau_j = \sum_{k \in I} \alpha_k \frac{1}{2\pi i} d \ln \frac{f_j}{f_k}.$$

We obtain

$$\begin{aligned} \kappa &= \sum_{j \in I_0} \int_{U_j} \bar{\partial} \tau_j = \sum_{j \in I_0} \int_{\partial U_j} \sum_{k \in I} \alpha_k \frac{1}{2\pi i} d \ln \frac{f_j}{f_k} \\ &= \sum_{j \in I_0} \frac{1}{2\pi i} \int_{\partial U_j} d \ln f_j - \sum_{j \in I_0} \sum_{k \in I} \frac{1}{2\pi i} \int_{U_k \cap U_j} \alpha_k d \ln f_k. \end{aligned} \tag{17.19}$$

Since the cochain $\{f_i\}$ is different from zero on the ideal boundary, the second addendum of the right-hand side of (17.19) may be evaluated as

$$\begin{aligned} \Delta &= \Big| \sum_{j \in I_0} \sum_{k \in I} \frac{1}{2\pi i} \int_{U_k \cap \partial U_j} \alpha_k d \ln f_k \Big| \\ &= \lim_{n \to \infty} \Big| \sum_{j \in I_{0n}} \sum_{k \in I} \frac{1}{2\pi i} \int_{U_k \cap \partial U_j} \alpha_k d \ln f_k \Big| \\ &= \lim_{n \to \infty} \Big| \int_{\partial M_n} \sum_{k \in I} \alpha_k d \ln f_k \Big| < \frac{1}{\varepsilon} \lim_{n \to \infty} \int_{\partial M_n} \sum_k |\alpha_k| |df_k|. \end{aligned}$$

Here $I_{0n} \subset I_0$ is the indices set determining a triangulation of M_n. The form $\nu = \sum_k |\alpha_k||df_k|$ has a finite Dirichlet integral, since

$$\begin{aligned} \|\nu\|_D^2 &\leqslant N \int_M \sum_{k \in I} |\alpha_k|^2 |df_k|^2 \\ &\leqslant N \sum_{k \in I} \int_{U_k} |df_k|^2 = N \|\{f_k\}\|_D^2. \end{aligned}$$

From which it follows that

$$\lim_{n\to\infty}\int_{\partial M_n}\nu = 0.$$

Therefore, we have

$$\kappa = \sum_{j\in I_0}\frac{1}{2\pi i}\int_{\partial U_j} d\ln f_j. \tag{17.19'}$$

By the condition of the theorem, only a finite number of addendums of the right-hand side of (17.19′) are different from zero, from which follows the statement of the theorem.

A solution of the problem (17.1) of the finite index is called strong if it is different from zero on the ideal boundary.

Let $\psi = \{\psi_i\} \in Cn_D^0(Q)$ be a finite positive divisor of the degree κ, a cochain $\{f_i'\} \in Cn_D^0(Q^*)$ and the cocycle $\{g_{ij}\}$ determining the bundle B_G be equal to

$$g_{ij} = f_i'\psi_{ij}f_j'^{-1}, \qquad \psi_{ij} = \psi_i/\psi_j. \tag{17.20}$$

Then the cochain $\{f_i\}$, $f_i = f_i'\psi_i$ is called a weak solution of the problem (17.1).

THEOREM 17.7. *A strong solution of the problem (17.1) of a finite index κ is also a weak solution.*

Indeed, by Theorem 17.6 the strong solution $\{f_i\}$ has κ zeros. Let these zeros be determined by a finite divisor $\{\psi_i\}$. Then the cochain $\{f_i'\}$, $f_i' = f_i/\psi_i$ is different from zero on M and on the ideal boundary of M and satisfies the relation (17.20). We estimate the norm

$$\begin{aligned}\|\{f_i'\}\|^2_{Cn_D^0(Q^*)} &= \sum_{i\in I}\int_{U_i}|d\ln f_i'|^2\\ &= \sum{}'\int_{U_i}|d\ln f_i'|^2\\ &\quad+\sum{}''\int_{U_i}|d\ln f_i'|^2.\end{aligned}$$

The first sum contains the finite number of addendums corresponding to the domains U_i in which the cochain $\{\psi_i\}$ is different from unity. This sum is finite. The second sum is formed by the domains in which $\psi_i \equiv 1$ and $f_i' = f_i$.

$$\sum{}''\int_{U_i}|d\ln f_i'|^2 \leqslant \frac{1}{\varepsilon^2}\sum{}''\int_{U_i}|df_i|^2 \leqslant \frac{1}{\varepsilon^2}\sum_{i\in I}\int_{U_i}|df_i|^2 = \frac{1}{\varepsilon^2}\|\{f_i\}\|_{Cn_D^0(Q)},$$

Q.E.D

Let $\{f_i\}$ be a weak solution of (17.1), the finite cochain $\{\psi_i\} \in Cn_D^0(Q)$ determine the divisor of zeros of the solution, and let $\bar{g}$ and $\bar{\psi}$ be the cohomology classes of the group $H_D^1(Q^*)$ determined by the cocycles $\{g_{ij}\}$and$\{\psi_{ij}\}$, respectively. Then, by (17.20), these classes coincide.

Consider the exact sequence

$$0 \longrightarrow J \longrightarrow Q \xrightarrow{e} Q^* \longrightarrow 0, \tag{17.21}$$

where $e(f) = \exp 2\pi i f(p)$. We obtain the exact D-cohomology sequence

$$H^1(J) \xrightarrow{i^*} H_D^1(Q) \xrightarrow{e^*} H_D^1(Q^*) \longrightarrow H^2(J) \longrightarrow \dots,$$

and the exact sequence

$$H^1(Q^*) \longrightarrow H^2(J) \longrightarrow H^2(Q).$$

Since $H^1(Q^*) = H^2(Q) = 0$ (Theorem 14.5), $H^2(J) = 0$, and we have the exact sequence

$$H^1(J) \xrightarrow{i} H_D^1(Q) \xrightarrow{e^*} H_D^1(Q^*) \longrightarrow 0.$$

Hence e^* is an epimorphism. Therefore, if the cocycles $\{g_{ij}\}$ and $\{\psi_{ij}\}$ are cohomological, then

$$e^{*-1}(\bar{g} - \bar{\psi}) \in i^* H^1(J),$$

and hence

$$e^{*-1}(\bar{g}) - e^{*-1}(\bar{\psi}) \in i^* H^1(J).$$

We obtain the following result.

THEOREM 17.8. *For a weak solvability of the Riemann problem (17.1) of the index $\kappa < \infty$, it is necessary and sufficient that there exist a finite positive divisor $\{\psi_i\} \in Cn_D^0(Q)$ of the degree κ such that the cohomology classes $e^{*-1}(\bar{g})$ and $e^{*-1}(\bar{\psi})$ coincide in the group $H_D^1(Q)/i^* H^1(J)$. This condition is necessary for the strong solvability of (17.1).*

It is natural to denote by

$$J_D = H_D^1(Q)/i^* H^1(J)$$

the D-Jacobian of the surface M. The subset of the D-Jacobian

$$J_S = S_1/i^* H^1(J)$$

has the structure of an infinite-dimensional torus, since the group

$$S_1 = H_D^0(A^{0,1})/\bar{\partial}^* H_D^0(A^0) \cong H_D^0(Q^1)$$

has the structure of the Hilbert space, and $i^* H^1(J)$ is a lattice. This torus is called the S-Jacobian of M. In the following section we shall show that for the S-Jacobian the solvability of the Riemann problem is equivalent to the problem of the inversion of Abelian integrals.

D. *The Case $\kappa = 0$. S-Problems*

The following theorem is a simple corollary of Theorem 17.8.

THEOREM 17.8′. *In order that the Riemann problem of the zero index have a weak holomorphic solution, it is necessary and sufficient that the cocycle* $\{g_{ij}\}$ *determine the zero cohomology class* $\bar{g}$ *of the group* $H^1_D(Q^*)$ *that is equivalent to the condition* $e^{*-1}(\bar{g}) \in i^* H^1(J)$.

The idea of S-divisors suggests the defining of S-problems whose solvability conditions relate to the S-Jacobian and can be studied in detail.

Consider the special covering (§16). Let h_n be point divisors in zones W_n whose points do not belong to any domains Δ_{n-1}, Δ_n,

(17.22)

$$h_n = \sum_k \alpha_{nk} p_{nk}, \quad n \geqslant 0; \quad \deg h_n = 0,$$

$$n \geqslant 1; \quad \deg h_0 = \kappa_0, \quad h_n \in V_n \backslash \Delta_n.$$

Assume

$$d\Omega_n = \sum_k \alpha_{nk} d\Omega_{p_n p_{nk}}(q), \qquad n \geqslant 0, \tag{17.23}$$

where p_n is an arbitrary fixed point belonging to $W_n \backslash (\Delta_{n-1} \cup \Delta_n)$ and $d\Omega_{p_i p_j}$ is the Abelian differential of the third kind on M (see §18A).

We assume in W_n that

$$\begin{aligned} h_n(q) &= \exp\{\Omega_n(q) + H_n(q) + H_n^0(q)\}, \qquad n \geqslant 1, \\ h_0(q) &= m_0(q) \exp\{\Omega_0(q) + H_0(q) + H_0^0(q)\}, \end{aligned} \tag{17.24}$$

where $\{H_n\} \in Cn^0(Q), \delta^*\{H_n\} \in S_1$. The functions $\Omega_n(q) + H_n^0(q)$ are single-valued in $M_n, n \geqslant 0, \{dH_n^0\} \in Cn_D^0(Q^1)$, and the functions $H_n^0(q)$ are holomorphic in M_n, and $m_0(q)$ is a function analytic in M_0 and different from zero, satisfying the boundary condition

$$m_0^+(p) = G(p) m_0^-(p), \qquad p \in L \subset M_0. \tag{17.25}$$

The cochain (17.24) determines some boundary problem. We suppose that the series

(17.26)

$$|R[dw]| = |\sum_{n=0}^{\infty} \left(\sum_k \alpha_{nk} \int_{p_k}^{p_{nk}} dw \right)| < C\|dw\|_{L_2(M)}, \qquad dw \in H_D^0(Q^1),$$

i.e., the functional $R[dw]$ is bounded on $H_D^0(Q^1)$.

Definition. A Riemann S-problem is the problem determined by the cochain (17.24) with the condition (17.26).

THEOREM 17.9. *The Cauchy index of an S-problem is equal to zero. The cocycle* $\{\frac{1}{2\pi i} \ln h_{n,n+1}\}$ *determined by an S-problem corresponds to the cohomology class* $\tilde{g} \in S_1 \subset H^1_D(Q)$.

PROOF: Construct the cochain (17.18). By virtue of the definition of the special covering, the terms $h_{n,n+1}$ in the domains Δ_n are the single terms different from unity. Taking into account (16.12), we have

$$\tau_n(q) = \begin{cases} \frac{1}{2\pi i} \alpha_{n+1}(q) d \ln h_n(q)/h_{n+1}(q), & q \in \Delta_n \\ 0 & q \in \Delta_n. \end{cases} \tag{17.27}$$

By (17.24) and (16.11′) we see that that

$$\kappa = \sum_{n=0}^{\infty}\int_{V_n}\bar\partial\tau_n = \sum_{n=0}^{\infty}\int_{\Delta_n}\bar\partial[\alpha_{n+1}\partial\ln(h_n/h_{n+1})]$$
$$= \sum_{n=0}^{\infty}\{\frac{1}{2\pi i}\int_{L_n}\alpha_{n+1}d\ln(h_n/h_{n+1}) - \frac{1}{2\pi i}\int_{L'_n}\alpha_{n+1}d\ln(h_n/h_{n+1})\}$$
$$= \sum_{n=0}^{\infty}\frac{1}{2\pi i}\int_{L_n}d\ln(h_n/h_{n+1}) = \sum_{n=1}^{\infty}\frac{1}{2\pi i}\int_{L_n-L_{n-1}}d\ln h_n + \frac{1}{2\pi i}\int_{L_0}d\ln h_0.$$

The integrals

$$\frac{1}{2\pi i}\int_{L_0}d\ln h_0 = \frac{1}{2\pi i}\int_{L_n}d\ln h_n - \frac{1}{2\pi i}\int_{L_{n-1}}d\ln h_n = 0,$$

since in every domain V_n a single-valued branch of the function $\ln h_n$ may be chosen. This implies the first statement of the theorem.

Now we shall show that the cocycle $\{\frac{1}{2\pi i}\ln h_{n,n+1}\}$ determines an element of the group S_1. Suppose that in the cochain (17.24) all $H_n = 0$, and assume that

$$T_n(q) = \frac{1}{2\pi i}\ln h_n(q) - \frac{1}{2\pi i}\sum_m(\alpha_m + \alpha'_m)\ln h_m(q), \tag{17.28}$$

where summation is done over all zones of the covering containing a point q. Let

$$s(q) = \frac{1}{2\pi i}\sum_m(\alpha_m + \alpha'_m)\ln h_m(q)$$

and calculate the values of the functional on $H^0_D(Q^1)$,
(17.29)

$$\tilde g[dw] = \int_M\bar\partial T_n\wedge dw = \sum_{n=0}^{\infty}\int_{V_n}\bar\partial T_n\wedge dw = \sum_{n=0}^{\infty}\frac{1}{2\pi i}\int_{V_n}\bar\partial(\ln h_n - s)\wedge dw$$
$$= \sum_{n=0}^{\infty}\frac{1}{2\pi i}\int_{\partial V_n}(\ln h_n - s)dw, \qquad dw\in H^0_D(Q^1).$$

It is easy to verify that

$$\begin{aligned}&\frac{1}{2\pi i}\int_{\partial V_n}\ln h_n dw = -\sum_k\alpha_{nk}\int_{p_n}^{p_{nk}}dw, \qquad n\geqslant 1,\\ &\frac{1}{2\pi i}\int_{\partial V_0}\ln h_0 dw = -\sum_k\alpha_{0k}\int_{p_0}^{p_{0k}}dw - \frac{1}{2\pi i}\int_L\ln G dw.\end{aligned} \tag{17.30}$$

On the other hand, taking into account (16.11′), we obtain

$$\sum_{n=0}^{\infty} \int_{\partial V_n} s\,dw = \lim_{n\to\infty} \int_{\partial M_n} s\,dw = \lim_{n\to\infty} \int_{L_n} \ln h_{n+1} dw = 0, \tag{17.31}$$

since the function $\ln h_{n+1}$ is holomorphic in M_n.

If the cochain $\{H_n\}$ is different from zero, it is represented in the form

$$H_n = f + t_n, \qquad \{t_n\} \in Cn^0_D(A^0),$$

and the functional $g[dw]$ has the added term

$$t[dw] = \sum_{n=0}^{\infty} \int_{V_n} \bar\partial t_n \wedge dw. \tag{17.32}$$

Taking into account (17.29), (17.30), and (17.31), we obtain

$$\tilde g[dw] = -\sum_{n=0}^{\infty} \Big(\sum_k \alpha_{nk} \int_{p_n}^{p_{nk}} dw\Big) - \frac{1}{2\pi i} \int_L \ln G\,dw + t[dw]. \tag{17.33}$$

From (17.26) all these values are finite. Therefore, the cocycle $\{\frac{1}{2\pi i} d \ln g_{n,n+1}\}$ generates a bounded linear functional on $H^0_D(Q^1)$ and hence generates the cohomology class $\tilde g \in S_1$.

THEOREM 17.10. *In order that a Riemann S-problem be weak solvable it is necessary and sufficient that for some exhaustion $M_n \searrow M$ the system*

$$\sum_{n=0}^{\infty}\Big(\sum_k \alpha_{nk} \int_{p_n}^{p_{nk}} dw_j\Big) - \frac{1}{2\pi i}\int_L \ln G\,dw_j + t[dw_j] \tag{17.34}$$

$$= \lim_{n\to\infty} \sum_{\mu=1}^{g_{n_k}} \Big(c_{2\mu-1}\int_{k_{2\mu}} dw_j - c_{2\mu}\int_{k_{2\mu-1}} dw_j\Big), \qquad j = 1, 2, \ldots$$

be solvable in integers. Here $dw_1, dw_2, \ldots$ is some basis of the space $H^0_D(Q^1)$.

From 14.11 there exists a real harmonic differential η_G, such that $\tilde g[dw] = -(dw, \eta_G)$. In order that the problem be solvable it is necessary and sufficient that all periods of the differentials η_G be integers. Because of Theorem 14.10, the functional considered can be represented in the form

$$\tilde g[dw] = \lim_{n_k\to\infty} \sum_{\mu=1}^{g_{n_k}} \Big\{ \int_{k_{2\mu-1}} \eta_G \int_{k_{2\mu}} dw - \int_{k_{2\mu}} \eta_G \int_{k_{2\mu-1}} dw \Big\}. \tag{17.35}$$

Equations (17.35) and (17.33) involve the statement of the theorem.

E. *Inversion of Abelian Integrals*

Return to the case of an arbitrary index. We can prove that for the Jacobian J_S the conditions of Theorem 17.8 mean the solvability of some problem of the inversion of Abelian integrals which analog for the compact surfaces is called the Jacobi inversion problem.

THEOREM 17.11. *If $e^{*-1}(\bar{g}) \in S_1$, then for the Riemann problem of a finite index to be weak solvable, it is necessary and sufficient that there exist a positive finite divisor $\psi = \sum_k \kappa_k q_k$, $\deg \psi = \sum_k \kappa_k = \kappa$, for which the system*

$$(17.36) \quad \sum_k \kappa_k w_j(q_k) = -\tilde{g}[dw_j] + \lim_{n_k \to \infty} \sum_{\mu=1}^{g_{n_k}} \Big(c_{2\mu-1} \int_{k_{2\mu}} dw_j - c_{2\mu} \int_{k_{2\mu-1}} dw_j\Big), \qquad j = 1, 2, \ldots$$

be solvable in integers. The values $\tilde{g}[dw_j]$ are determined by the relation (17.33).

PROOF: From Theorem 17.8 the necessary and sufficient condition of the solvability of the problem is the existence of the positive divisor ψ, $\deg \psi = \kappa$, such that

$$\bar{h}_0 = e^{*-1}(\bar{\psi}) - e^{*-1}(\bar{g}) \in i^* H^1(J).$$

Because of the conditions of the theorem, the cocycle $e^{*-1}(\bar{g})$ corresponds to the functional on $H_D^0(Q^1)$. The cocycle $\{\psi_{ij}\}$ is finite and hence $e^{*-1}(\bar{\psi}) \in S_1$. Therefore, $\{\psi_{ij}\}$ corresponds to the functional on $H_D^0(Q^1)$. The same is valid for the class $\bar{h}_0$. The functional $\tilde{g}$ associated with the cocycle g is represented by the formula (17.33). The element $\bar{h}_0$ corresponds to the functional $\tilde{h}_0[dw]$ which is represented by

$$(17.37) \qquad \tilde{h}_0[dw] = \lim_{n_k \to \infty} \sum_{\mu=1}^{g_{n_k}} \Big(c_{2\mu-1} \int_{k_{2\mu-1}} dw - c_{2\mu} \int_{k_{2\mu}} dw\Big)$$

(see (17.35)). This class $\bar{h}_0$ belongs to $i^* H^1(J)$ if and only if all numbers c_j are integers. The functional $\tilde{\psi}[dw]$ corresponding to the finite divisor $\psi = \sum_k \kappa_k q_k$ is calculated analogously to (17.33) and is equal to

$$(17.38) \qquad \tilde{\psi}[dw] = \sum_k \kappa_k \int_{q_0}^{q_k} dw.$$

Eq. (17.37) and Eq. (17.38) involve (17.36).

§18 The Solving of the Riemann Problem in the Explicit Form

A. *Cauchy-type Integrals*

Let α be a real closed differential with a finite integral Dirichlet. From the Hodge–Royden theorem (14.2) it follows that there exists the harmonic differential ϕ with a finite Dirichlet integral whose periods coincide with the periods of α. Because the differential $*\varphi$ also has a finite Dirichlet integral, we obtain the Abelian differential of the first kind with a finite Dirichlet integral whose periods have imaginary parts which are equal to periods of α. In particular, we use the basis of the space $H_D^0(Q^1)$ (see(2.13)),

$$\begin{aligned}(18.1)\qquad &\operatorname{Im}\int_{k_{2\mu}} d\theta_{2\nu-1} = -\delta_{\mu\nu}, \qquad \operatorname{Im}\int_{k_{2\mu-1}} d\theta_{2\nu-1} = 0,\\ &\operatorname{Im}\int_{k_{2\mu-1}} d\theta_{2\nu} = \delta_{\mu\nu}, \qquad \operatorname{Im}\int_{k_{2\mu}} d\theta_{2\nu} = 0,\\ &\mu,\nu = 1,2,\ldots.\end{aligned}$$

On a Riemann surface M with zero boundary, there exists an Abelian differential of the third kind, $d\Omega_{q_0q}(p)$, with poles with residues ∓ 1 at the points q_0, q and a finite Dirichlet integral

$$(18.2)\qquad \int_{M\setminus\Delta} |d\Omega_{q_0q}(p)|^2 < \infty,$$

where Δ is some neighbourhood of the points q_0, q and periods (see (2.16), (3.12))

$$(18.3)\qquad \int_{k_j} d\Omega_{q_0q} = 2\pi i \operatorname{Im}\int_{q_0}^{q} d\theta_j, \qquad j = 1,2,\ldots.$$

We use the Cauchy kernel (see (4.4))

$$(18.4)\qquad M_*(p,q)dz(p) = \partial_p[\Omega_{p_0p}(q) - \Omega_{p_0p}(q_0)].$$

It is clear that periods of the kernel are (see (4.7))

$$(18.5)\qquad L_j(p) = \int_{k_j} d_q M_*(p,q)dz(p) = \partial_p \operatorname{Im}\int_{p_0}^{p} d\theta_j = \pi d\theta_j(p).$$

Consider the Cauchy type integral

$$(18.6)\qquad f(q) = \frac{1}{2\pi i}\int_L \varphi(p)M_*(p,q)dz(p),$$

where L is a compact closed Liapounov contour and $\varphi(p)$ is a Hölder continuous function. We suppose that the contour L separates the surface into two domains $T^{\pm}$ (these domains can be noncompact).

We have, $s \in L$,

$$\begin{aligned}(18.7)\qquad f^+(s) &= \frac{1}{2}\varphi(s) + \frac{1}{2\pi i}\int_L \varphi(p)M_*(p,s)dz(p),\\ f^-(s) &= -\frac{1}{2}\varphi(s) + \frac{1}{2\pi i}\int_L \varphi(p)M_*(p,s)dz(p),\end{aligned}$$

periods $f(q)$ are equal to

$$f_j = \int_{k_j} df = \frac{1}{2i} \int_L \varphi(p) d\theta_j, \qquad j = 1, 2, \ldots. \tag{18.8}$$

By virtue of (18.2) a Dirichlet integral of $f(q)$ is finite.

B. *Construction of a Solution*

In this section, we shall study an S-problem determined by the cochain (see (17.24))

$$\begin{aligned} h_0(q) &= m_0(q) \exp[\Omega_0(q) + H_0^0(q)], \\ h_n(q) &= \exp[\Omega_n(q) + H_n^0(q)], \qquad n \geqslant 1 \end{aligned} \tag{18.9}$$

in zones V_n, W_n. Here the functions $\Omega_n(q) + H_n^0(q)$ are single-valued analytic functions in $M_n, n \geqslant 0, \{dH_n^0\} \in Cn_D^0(Q^1), \{H_n^0\} \in Cn^0(Q)$, and

$$d\Omega_n = \sum_k \alpha_{nk} d\Omega_{p_n p_{nk}}(q), \qquad n \geqslant 0. \tag{18.10}$$

Using Eq. (18.3), we obtain the relations

$$\int_{k_j} d\Omega_n = 2\pi i \operatorname{Im} \sum_k \int_{p_n}^{p_{nk}} d\theta_j, \qquad j = 1, 2, \ldots. \tag{18.11}$$

We assume that the series

$$d\Omega = \sum_{n=0}^{\infty} d\Omega_n(q) \tag{18.12}$$

converges uniformly in every closed subdomain of M containing no singularities of the differentials $d\Omega_n (n = 0, 1, \ldots)$. By virtue of Eq. (18.3) the periods of $d\Omega$ along the cycles $k_\nu (\nu = 1, 2, \ldots)$ are equal to

$$\Omega_\nu = \int_{k_\nu} d\Omega = 2\pi i \sum_{n=0}^{\infty} \left(\sum_k \alpha_{nk} \operatorname{Im} \int_{p_n}^{p_{nk}} d\theta_\nu \right), \qquad \nu = 1, 2, \ldots. \tag{18.13}$$

Let the integers κ_j be determined by the relations

$$\kappa_j = \frac{1}{2\pi} \Delta_{L_j} \arg G, \qquad \kappa_0 = \operatorname{ind}_L G = \sum_{j=1}^m \kappa_j,$$

where L_j are connected components of L. On every curve L_j, we choose the point p_j and count the value $\ln G(p)$ from p_j. At this point the function $\ln G(p)$ has a jump which is equal to $2\pi i \kappa_j$. Consider the Cauchy-type integral

$$f(q) = \frac{1}{2\pi i} \int_L \ln G(p) M_*(p, q) dz(p). \tag{18.14}$$

Periods of the function along the cycles k_j are equal to

$$(18.15)\qquad f_j = \frac{1}{2i}\int_L \ln G d\theta_j, \qquad j = 1, 2, \ldots .$$

and the Dirichlet integral is finite.

Consider the function

$$(18.16)\qquad R(q) = \exp\{f(q) + \Omega(q)\},$$

$f(q)$ and $\Omega(q)$ are determined by Eq. (18.12) and (18.14). This function is a weak multiple-valued solution of the problem (17.1) with multiplicative periods $\exp(f_j + \Omega_j)$ along the cycles k_j.

Let $h = \sum_k \delta_k q_k \geqslant 0$ be an arbitrary finite divisor of the degree $\kappa, \deg h = \kappa, h \in M_0$. Assume

$$(18.17)\qquad d\tilde{\Omega} = \sum_k \delta_k d\Omega_{p_0 q_k} - \sum_l \kappa_l d\Omega_{p_0 s_l}.$$

By virtue of (18.3) and (18.12), the periods of this integral are equal to

$$(18.18)\qquad \tilde{\Omega}_j = \int_{k_j} d\tilde{\Omega} = 2\pi i \operatorname{Im}\left\{\sum_k \delta_k \theta_j(q_k) - \sum_l \kappa_l \theta(s_l)\right\}.$$

If there exists a single-valued solution of the problem (17.1), it is represented in the form

$$(18.19)\qquad F(q) = \exp[f(q) + \Omega(q) + \tilde{\Omega}(q) + Z(q)].$$

Here $Z(q)$ is an Abelian integral of the first kind whose periods Z_j along k_j satisfy the conditions

$$(18.20)\qquad f_j + \Omega_j + \tilde{\Omega}_j + Z_j = 2\pi i c_j, \qquad j = 1, 2, \ldots,$$

where c_j are integers. Therefore, $\operatorname{Re} Z_j$ are equal to $-\operatorname{Re} f_j$. As follows from (18.1),

$$(18.21)\qquad dZ = -i\sum_{\mu+1}^{\infty}(\operatorname{Re} f_{2\mu}\cdot d\theta_{2\mu-1} - \operatorname{Re} f_{2\mu-1}\cdot d\theta_{2\mu}).$$

We obtain from (18.20)

$$(18.22)\qquad \operatorname{Im} f_j + \operatorname{Im}\Omega + \operatorname{Im}\tilde{\Omega} + \operatorname{Im} Z_j = 2\pi c_j, \qquad j = 1, 2 \ldots .$$

If these conditions are satisfied, the cochain

$$(18.23)\qquad \varphi_n(q) = \exp\{-\Omega_n(q) - H_n^0(q)\}F(q)$$

determines a weak solution of the problem. If the cochain $\{\ln \varphi_n\}$ is bounded on the ideal boundary, we get a strong solution. If $\{\varphi_n\} \in Cn^0_D(Q)$, it is a solution in the sense of §17A.

Therefore, the Riemann problem is reduced to the system (18.22). Using Eq. (18.11), (18.12), (18.18), and (18.21), we obtain the following problem of inversion of Abelian integrals

(18.24)

$$2\pi \sum_k \delta_k \operatorname{Im} \theta_j(q_k) = 2\pi \sum_l \kappa_l \operatorname{Im} \theta_j(s_l) - \operatorname{Im} f_j - 2\pi \sum_{n=0}^{\infty} \Big(\sum_k \alpha_{nk} \int_{p_n}^{p_{nk}} d\theta_j \Big)$$
$$+ \sum_{\mu=1}^{\infty} \Big(\operatorname{Re} f_{2\mu} \cdot \operatorname{Re} \int_{k_j} d\theta_{2\mu-1} - \operatorname{Re} f_{2\mu-1} \cdot \operatorname{Re} \int_{k_j} d\theta_{2\mu} \Big) + 2\pi c_j.$$

Expand the differentials $id\theta_j$ into the series using the real basis $\{d\theta_j\}$. Using (18.1) we obtain

$$id\theta_j(q) = \sum_{\mu=1}^{\infty} \left\{ \Big(\operatorname{Re} \int_{k_j} d\theta_{2\mu-1}\Big) d\theta_{2\mu}(q) - \Big(\operatorname{Re} \int_{k_j} \delta\theta_{2\mu}\Big) d\theta_{2\mu-1}(q) \right\}.$$

From Eq. (18.15) it follows that

$$\operatorname{Im} f_j = \operatorname{Im} \frac{1}{2i} \int_l \ln G d\theta_j = -\operatorname{Re} \frac{1}{2i} \int_L \ln G(id\theta_j)$$

(18.25)

$$= -\sum_{\mu=1}^{\infty} \left\{ \Big(\operatorname{Re} \int_{kj} d\theta_{2\mu-1}\Big) \operatorname{Re} f_{2\mu} - \Big(\operatorname{Re} \int_{k_j} d\theta_{2\mu}\Big) \operatorname{Re} f_{2\mu-1} \right\}$$

Substituting Eq. (18.25) into Eq. (18.24) we get

(18.26)
$$2\pi \operatorname{Im} \sum_k \delta_k \theta_j(q_k) = 2\pi \operatorname{Im} \sum_l \kappa_l \theta_j(s_l) - 2\operatorname{Im} f_j$$
$$- 2\pi \sum_{n=0}^{\infty} \Big(\sum_k \alpha_{nk} \operatorname{Im} \int_{p_n}^{p_{nk}} d\theta_j \Big) + 2\pi c_j, \qquad j = 1, 2, \ldots.$$

Note that

(18.27)
$$c_j = \sum_{\mu=1}^{\infty} \Big(c_{2\mu} \operatorname{Im} \int_{k_{2\mu-1}} d\theta_j - c_{2\mu-1} \operatorname{Im} \int_{k_{2\mu}} d\theta_j \Big)$$

The relation (18.26) is valid for any Abelian differential of the first kind belonging to $H^0_D(Q^1)$ since $\{d\theta_j\}$ is a basis. Hence Eq. (18.26) is valid for the differentials $id\theta_j$. Therefore, this relation may be written in the form

(18.28)
$$\sum_k \delta_k \theta_j(q_k) = \sum_l \kappa_l \theta_j(s_l) - \frac{1}{2\pi i} \int_L \ln G(p) d\theta_j(p)$$
$$- \sum_{n=0}^{\infty} \Big(\sum_k \alpha_{nk} \int_{p_n}^{p_{nk}} d\theta_j \Big) + \sum_{\mu=1}^{\infty} \Big(c_{2\mu} \int_{k_{2\mu-1}} d\theta_j - c_{2\mu-1} \int_{k_{2\mu}} d\theta_j \Big).$$

where c_j are integers, $(j = 1, 2, \ldots)$.

CHAPTER 5

GENERALIZED ANALYTIC FUNCTIONS

§19 Bers–Vekua Integral Representations

A. *Generalized Analytic Functions on a Plane*

On a complex plane, we consider the system of differential equations

$$\begin{aligned} \xi_x - \eta_y &= a\xi + b\eta \\ \xi_y + \eta_x &= a_1\xi + b_1\eta. \end{aligned} \tag{19.1}$$

This system is written in the complex form

$$\begin{aligned} \bar\partial u &= Au + B\bar u, & u &= \xi + i\eta, \\ \bar\partial &= \frac{1}{2}\left(\frac{\partial}{\partial x} + i\frac{\partial}{\partial y}\right), & x &= x + iy. \end{aligned} \tag{19.2}$$

One calls the system (19.1), (19.2) the Carleman system or the Bers–Vekua system. We prefer to call this system the system of Carleman–Bers–Vekua (CBV). Solutions to this system are called *generalized analytic functions* or *pseudo-analytic functions.* The generalized analytic function theory was founded in the 1950s by Bers [1], [a] and Vekua [35], [36].

In the 1960s the CBV system was studied for matrix coefficients. This theory is close to the general problem of the index of elliptic operators (see the fundamental monograph by Wendland [37]).

Generalized analytic functions on Riemann surfaces were first studied by Bers [b]. Then, simultaneously, the paper of Koppelman [c] and the series of the author's papers [h–l] appeared. All of these papers form the subject of this chapter. The author's monograph [30] should also be pointed out.

If $A(z), B(z)$ are smooth functions in the bounded domain T, we have two Bers–Vekua representations for solutions of (19.2),

$$u(z) = \varphi(z)\exp -\frac{1}{\pi}\iint\limits_T [A(t) + B(t)\frac{\overline{u(t)}}{u(t)}]\frac{d\sigma_t}{t-z}, \tag{19.3}$$

$$u(z) + \frac{1}{\pi}\iint_T [A(t)u(t) + B(t)\overline{u(t)}]\frac{d\sigma_t}{t-z} = F(z), \tag{19.4}$$

where $\varphi(z)$ and $F(z)$ are analytic functions and $d\sigma_t = \frac{1}{2i}d\bar{t}\wedge dt$.

Taking into account that the integral

$$(Tf)(z) = -\frac{1}{\pi}\iint_T f(t)\frac{d\sigma_t}{t-z} \tag{19.5}$$

has the property (see (9.6))

$$\bar{\partial}(Tf) = \begin{cases} f(z) & z \in T \\ 0 & z \in E\backslash T, \end{cases} \tag{19.6}$$

E is the complex plane, and we can verify the representations (19.3)–(19.4) by differentiation.

Equation (19.3) means that zeros and singularities of the function $u(z)$ are determined by the analytic function $\varphi(z)$. The argument also follows from (19.3).

Equation (19.4) is solvable for any right-hand side. In fact, a solution of the homogeneous equation (19.4) is a generalized analytic function in T, analytic in $E-T$, and has zero at infinity. As follows from the argument principle, such a function is zero.

If $A, B \in L_q(T), q > 2$, the representations (19.3)–(19.4) are also valid. The operator T is studied as an operator of the potential type [36]. We have the following theorem for the operator

$$(Pf)(z) = -\frac{1}{\pi}\iint_T [A(t)f(t) + B(t)\overline{f(t)}]\frac{d\sigma_t}{t-z}. \tag{19.7}$$

THEOREM 19.1. *([36], Theorems 1.29, 1.35, 1.25) If $A, B \in L_q(T) q > 2$, then the function (19.7) is analytic in E–T, and the operator P determines the completely continuous mapping $P : L_\alpha(T) \longrightarrow L_\gamma(T), \frac{1}{2} \leqslant \frac{1}{\alpha} + \frac{1}{q} \leqslant 1, \frac{1}{\alpha} + \frac{1}{q} - \frac{1}{2} < \frac{1}{\gamma} < 1$. The estimate*

$$\|Pf\|_{L_\gamma(T)} \leqslant K\|f\|_{L_\alpha(T)} \tag{19.8}$$

is valid. In the domain T the function $(Pf)(z)$ has the generalized derivations and $\bar{\partial}(Pf) = Af + B\bar{f}$. For $\gamma > 2q/(q-2)$, the function $(Pf)(z)$ is Hölder continuous.

It is clear that the solutions of Eq. (19.2) are treated in the general sense (for exact definitons, see [36]).

B. *Generalized Analytic Functions on a Compact Riemann Surface.*

Basic Definitions

In order for Eq. (19.2) to be invariant on a Riemann surface, it is necessary that $A(p)\overline{dz(p)}, B(p)\overline{dz(p)}$ be forms of the type (1,0), i.e., that these coefficients depend on the local coordinates by the rule

$$A(z^*(p)) = A(z(p))\frac{\overline{dz(p)}}{\overline{dz^*(p)}}, \quad B(z^*(p)) = B(z(p))\frac{\overline{dz(p)}}{\overline{dz^*(p)}}. \tag{19.9}$$

Fix some triangulation $\{U_i\}$ of the surface M and local coordinates $z_i(p)$ in U_i. Define the spaces $L_p^{\alpha,\beta}(M)$ of the forms of the type $(\alpha,\beta)(\alpha,\beta = 0,1)$ by the norms

$$\|\omega\|_{L_p^{(\alpha,\beta)}(M)} = \sum_i \|\omega\|_{L_p^{(\alpha,\beta)}}(U_i). \tag{19.10}$$

It is clear that passing from one triangulation and coordinate system to another, we change the norm (19.10) on a topology equivalent.

We shall study the CBV system (19.2) on the compact surface M for $A, B \in L_q^{0,1}(M), q > 2$. Solutions are treated for the general case. A solution has a pole at some point $p \in M$ if the function $\varphi(z)$ in (9.3) has a pole of the corresponding order.

Let $\widehat{M}$ be a surface M cut along the cycles $k_1, \dots, k_{2g}$ of the homology basis. Parallel with single-valued solutions of Eq. (19.2), we shall consider multivalued functions whose fixed branches on $\widehat{M}$ satisfy Eq. (19.2). Functions obtaining additive increments in tracing of cycles k_j are called *integrals*. Functions of the type $\exp H(q)$, where $H(q)$ is an integral are called *multipliers*.

We shall also consider differential forms of the type (1,0) $v(q)dz(p)$, satisfying (in the general sense) the equation

$$\bar\partial v + Av + \overline{Bv} = 0. \tag{19.11}$$

Consider the multivalued Cauchy kernel $M^*(p,q)dz(p)$ (4.3) and the operator

$$Pf = -\frac{1}{\pi}\iint\limits_{\widehat{M}} \left[A(p)f(p) + B(p)\overline{f(p)}\right] M^*(p,q)d\sigma_p. \tag{19.12}$$

The kernel branch on $\widehat{M}$ is determined by the condition $M^*(p,q_0) = 0$ (see §4). Here $f(p)$ is a single-valued function on M or some branch of the multivalued function (such a branch has jumps on the cuttings $k_1, \dots, k_{2g}$). The function $(Pf)(q)$ is multivalued on M and on account of (4.7) has periods

$$(Pf)_\mu = \int\limits_{k_{2\mu}} d(Pf) = -2i\iint\limits_{\widehat{M}} \left[A(p)f(p) + B(p)\overline{f(p)}\right] w'_\mu(p)d\sigma_p, \quad \int\limits_{k_{2\mu-1}} d(Pf) = 0, \qquad \mu = 1, \dots, g. \tag{19.13}$$

From Theorem 19.1 and Eq. (19.10) follows the folowing theorem.

THEOREM 19.2. *Operator (19.12) determines the completely continuous mapping*

$$P : L^0_\alpha(M) \longrightarrow L^0_\gamma(M), \qquad \frac{1}{2} \leqslant \frac{1}{\alpha} + \frac{1}{q} < 1, \qquad \frac{1}{q} + \frac{1}{\alpha} - \frac{1}{2} < \frac{1}{\gamma} < 1.$$

The estimate

$$\|Pf\|_{L^0_\gamma(M)} \leqslant K\|f\|_{L^0_\alpha(M)} \tag{19.14}$$

is valid. The function $(Pf)(q)$ has generalized derivations and

$$\bar{\partial}(Pf) = Af + B\bar{f}.$$

For $\gamma > 2q/(q-2)$, the function $(Pf)(z)$ is Hölder continuous. Theorem 19.2 involves two Bers–Verua representations.

THEOREM 19.3. *If $u(q)$ is a generalized analytic integral whose pole orders are no more than one, then the relation*

$$Ku \equiv u(q) + \frac{1}{\pi} \iint\limits_{\widehat{M}} \left[A(p)u(p) + B(p)\overline{u(p)}\right] M^*(p,q)\,d\sigma_p = F(q) \tag{19.15}$$

is valid. Here $F(q)$ is an Abelian integral. Periods of $u(q)$ along $k_{2\mu}$ are equal to

$$\begin{aligned} u_\mu &= F_\mu - 2i \iint\limits_{\widehat{M}} \left[A(p)u(p) + B(p)\overline{u(p)}\right] w\mu'(p)\,d\sigma_p, \\ F_\mu &= \int\limits_{k_{2\mu}} dF, \qquad \mu = 1, \ldots, g. \end{aligned} \tag{19.16}$$

Periods of $u(q)$ along $k_{2\mu-1}$ coincide with corresponding periods of $F(q)$.

THEOREM 19.4. *If $u(q)$ is a general analytic multiplier, then the representation*

$$u(q) = \exp\{\omega(q) - \frac{1}{\pi} \iint\limits_{\widehat{M}} \left[A(p) + B(p)\frac{\overline{u(p)}}{u(p)}\right] M^*(p,q)\,d\sigma_p\} \tag{19.17}$$

is valid. Here $\omega(q)$ is an Abelian integral of the first or the third kind.

Both representations are verified by differentiation. Equation (19.16) results from Eq. (19.13).

THEOREM 19.5. *If $v(q)dz(q)$ is a generalized analytic differential satisfying Eq. (19.11) and having no poles on M, the relation*

$$\widetilde{K}v = v(q) + \frac{1}{\pi} \iint\limits_{\widehat{M}} \left[A(p)v(p) + \overline{B(p)v(p)}\right] M^*(q,p)\,d\sigma_p = Z'(q) \tag{19.18}$$

is valid. Here $dZ = Z'dz$ is an Abelian differential of the first kind.

The representation (19.18) is verified by differentiation. As long as $v \in L_{\gamma}^{1,0}(M), \gamma > 2$ the right-hand side can have a single pole of the first order at $q = p_0$, stipulated by the pole of $M^*(p,q)$ (see §4). Then, because of the residues theorem, dZ is an Abelian differential of the first kind.

C. *The First Bers–Vekua Equation*

Consider the representation (19.15) as the integral equation with respect to $u(q)$. As was shown in Rodin [b], the corresponding homogeneous equation has, in general, nonzero solutions. If $F(q)$ has poles of no more than the first order, Eq. (19.15) may be considered in the space $L_{\alpha}^{0}(M), \frac{1}{2} \leqslant \frac{1}{\alpha} + \frac{1}{q} < 1$.

Define the bilinear form

$$(u, w) = \operatorname{Re} \iint_{\widehat{M}} u(p)\overline{w(p)}d\sigma_p,$$

where $wd\bar{z} \wedge dz$ is a form of the type (1,1). The operator K^* conjugate to (19.15) is

$$K^* w = \frac{1}{\pi} \iint_{\widehat{M}} \left[\overline{A(q)}w(p)\overline{M^*(q,p)} + B(q)\overline{w(p)}M^*(q,p)\right] d\sigma_p. \tag{19.19}$$

Therefore, in order for Eq. (19.15) to be solvable, it is necessary and sufficient that

$$\operatorname{Re} \iint_{\widehat{M}} F(q)\overline{w_j(q)}d\sigma_q = 0, \qquad j = 1, \ldots, s, \tag{19.20}$$

where $w_j (j = 1, \ldots, s)$ is a complete set of solutions of the homogeneous equation

$$w + K^* w = 0. \tag{19.21}$$

Let us assume that

$$v(q) = \frac{1}{\pi} \iint_{\widehat{M}} \overline{w(p)}M^*(q,p)d\sigma_p. \tag{19.22}$$

Then Eq. (19.21) involves

$$\overline{w(q)} = -A(q)v(q) - B(q)\overline{v(q)}.$$

Substituting this relation in (19.22), we obtain the equation

$$\widetilde{K}v = v(q) + \frac{1}{\pi} \iint_{\widehat{M}} [A(p)v(p) + \overline{B(p)v(p)}]M^*(q,p)d\sigma_p = 0. \tag{19.23}$$

The conditions (19.20) are rewritten in the form

$$< F, v_j >= 0, \qquad j = 1, \dots, s, \tag{19.24}$$

where

$$\begin{aligned} < u, v > &= \operatorname{Re} \iint\limits_{\widehat{M}} \left[A(p)u(p) + B(p)\overline{u(p)}\right] v(p) d\sigma_p \\ &= \operatorname{Re} \iint\limits_{\widehat{M}} u(p) \left[A(p)v(p) + \overline{B(p)v(p)}\right] d\sigma_p. \end{aligned} \tag{19.25}$$

Note. The operator $\widetilde{K}$ is conjugate to K with respect to the bilinear form (19.25). But this form can degenerate, and we prefer a stronger variant.

A solution of Eq. (19.23) $v(q)dz(q)$, is a generalized analytic differential satisfying Eq. (19.11).

Examine the nonhomogeneous equation

$$\widetilde{K}v = v(q) + \frac{1}{\pi} \iint\limits_{\widehat{M}} \left[A(p)v(p) + \overline{B(p)v(p)}\right] M^*(q,p) d\sigma_p = R(q), \tag{19.26}$$

where $Rdz(q)$ is a form of the type (1,0). Repeating our considerations, we become convinced of the following. In order that Eq. (19.26) be solvable in the space $L^{1,0}_{\gamma}(M)$, $\frac{1}{2} < \frac{1}{\gamma} + \frac{1}{q} < 1$, it is necessary and sufficient that

$$< u_j, R >= 0, \qquad j = 1, \dots, s, \tag{19.27}$$

where $u_j (j = 1, \dots, s)$ is a complete set of solutions of the equation $Ku = 0$.

D. *Equation* $\bar{\partial} u = Au$

We obtain the representation of solutions of the equation

$$\bar{\partial} u = Au \tag{19.28}$$

in the form

$$u(q) = \varphi(q) \exp -\frac{1}{\pi} \iint\limits_{\widehat{M}} A(p) M^*(p,q) d\sigma_p. \tag{19.29}$$

The second multiplier of the right-hand side of (19.29) has multiplicative periods along $k_{2\mu}$,

$$u_\mu = -2i \iint\limits_{\widehat{M}} A(p) w'_\mu(p) d\sigma_p. \tag{19.30}$$

Therefore the coefficient $A(p)$ leads to the shift of periods analogous to the factor

$$\exp \frac{1}{2\pi i} \int\limits_L \ln G(p) M^*(p,q) dz(p)$$

in the Riemann boundary problem (§5).

Hence Eq. (19.28) is equivalent to the Riemann boundary problem (see Rodin [f]). Thus, we have the following theorem.

THEOREM 19.6. *In order that there exist the solution of (19.28) determined by the divisor* $D = \sum \alpha_k p_k, \deg D = 0$, *it is necessary and sufficient that*

$$\sum \alpha_k w_\mu(p_k) \equiv \frac{1}{\pi} \iint\limits_{\widehat{M}} A(p) w'_\mu(p) d\sigma_p \qquad (\text{mod periods of } w_\mu) \tag{19.31}$$

(see §7).

Now we consider the case when $A(p)$ and $u(p)$ are $n \times n$ matrices.

LEMMA. *In the disk* $|z| < 1$ *there exists everywhere a regular nondegenerate matrix solution* $v(z)$ *of (19.28).*

Indeed, the equation (19.28) in the disk $|z| < 1$ is equivalent to the integral equation

$$Ku \equiv u(z) + \frac{1}{\pi} \iint\limits_{|t|<1} A(t) u(t) \frac{d\sigma_t}{t - z} = \Phi(z), \tag{19.32}$$

where $\Phi(z)$ is an analytic matrix. Let $u^1, \dots u^s$ be linear independent vector solutions of the homogeneous equation $Ku = 0$. Any solution $u(z)$ of $Ku = 0$ is analytically continuable into the domain $|z| > 1$ and $u(\infty) = 0$.

If $u(z)$ is a matrix solution of $Ku = 0$, then

$$\bar{\partial}(\det u) = SpA \det u. \tag{19.33}$$

Hence $\det u(z) \equiv 0$. Therefore, the number s of vector solutions of $Ku = 0$ is less than n, $s < n$.

Let $v = (v^1, \dots, v^n)$ be a matrix solution of the equation $Kv = P$, where P is a polynomial matrix, $Kv^j = P^j_m$. Here P^j_m are vectors whose components are polynomials of the degree m, and $\det v(z) \not\equiv 0$. The function $\det v(z)$ has the pole of degree $m_0 = nm$ at infinity and hence has m_0 zeros in the z-plane. Let z_1 be one of these zeros, $\det v(z_1) = 0$. Then there exist constants c_j such that

$$c_1 v^1(z_1) + \cdots + c_n v^n(z_1) = 0.$$

Let $c_j \neq 0$. Change the columns $v^j(z)$ of the matrix $v(z)$ by the column

$$v_1^j(z) = \frac{1}{z - z_1}(c_1 v^1(z) + \cdots + c_n v^n(z)). \tag{19.34}$$

From this, we obtain the matrix $v_1(z)$ whose determinant is equal to $\det v(z) = \det v(z)/(z - z_1)$. The function $\det v_1(z)$ has no more than $m_0 - 1$ zeros. Repeating this reseption m_0 times, we obtain the matrix solution $v(z)$, such that $\det v(z)$ is different from zero for $|z| < \infty$.

Return to the case of an arbitrary Riemann surface M. Let $\{U_i, i \in I\}$ be a covering of M. Then in any coordinate domain U_j a matrix solution $u(q)$ can be represented by

$$u(q) = v_j(q) \varphi_j(q), \tag{19.35}$$

where $v_j(q)$ is a nondegenerate matrix solution of (19.28) and $\varphi_j(q)$ is an analytic matrix. This means that the cochain $\{\varphi_j\}$ is a holomorphic section of the vector bundle B_A determined by the transition matrices (Röhrl [c]),

$$h_{jk}(q) = v_j(q)v_k^{-1}(q). \tag{19.36}$$

If T is a finite-connected domain belonging to M, then any vector bundle possesses a holomorphic nondegenerate matrix section over T (see Chapter 4 and [12]). Let a contour L separate the surfaces into domains T^+ and T^-, and let $v_0^\pm(q)$ be nondegenerate solutions of (19.28) in $T^\pm$, respectively. We obtain the representations

$$u(q) = v_0^\pm(q)\varphi_\pm(q), \qquad q \in T^\pm, \tag{19.37}$$

where $\varphi_\pm(q)$ are analytic in $T^\pm$. We obtain the matrix Riemann problem

$$\begin{aligned} \phi_+(p) &= G(p)\phi_-(p), \\ G(p) &= [v_0^+(p)]^{-1}v_0^-(p), \qquad p \in L, \end{aligned} \tag{19.38}$$

and the bundles B_G and B_A are equivalent. Conversely, any matrix Riemann problem is reduced to (19.28) if to continue continuously the matrix G into the domain T^- (Bojarskii [c]). We obtain

$$\begin{aligned} u(q) &= \Phi_+(q), & q \in T^+, \\ u(q) &= G(q)\Phi_-(q), & q \in T^-, \end{aligned}$$

$$\bar\partial u = Au \tag{19.39}$$

$$A(q) = \begin{cases} 0 & q \in T^+ \\ \bar\partial G \cdot G^{-1} & q \in T^- \end{cases}$$

§20 The Riemann–Roch Theorem

A. *Generalized Constants*

We call the solutions of Eq. (19.2) without poles *generalized constants*. The representation (19.17) involves the argument principle and the Liouville theorem. Hence, the real dimension of the space L_0 of generalized constants $l_0 \leqslant 2$. The examples are known as $l_0 = 0, 1, 2$ [30].

THEOREM 20.1. *Generalized constants are solutions of the equation $Ku = c$, $c \neq 0$ is a constant. Conversely, a single-valued solution of this equation is a generalized constant.*

Indeed, because of the argument principle, the generalized constant $u_0(q)$ has no zeros. Hence, Ku_0 is an Abelian integral of the first kind (see (19.16)) with zero periods along cycles $k_{2\mu-1}(\mu = 1, \ldots, g)$. Such an integral is a constant $Ku_0 = u_0(q_0) = c \neq 0$ because of the normalization of $M^*(p, q)$.

THEOREM 20.2. *Let $u_1, \dots, u_s$ be a basis of the space of solutions of the equation $Ku = 0$. Then assuming that $d\tilde{\omega}_{2\mu-1} = dw_\mu, d\tilde{\omega}_{2\mu} = idw_\mu, \mu = 1, \dots, g$, we obtain that*

$$\operatorname{rank} \| < u_i, \tilde{\omega}'_j > \| = s, \qquad i = 1, \dots, s, \quad j = 1, \dots, 2g. \tag{20.1}$$

Indeed, if $\operatorname{rank} \| < u_i, \tilde{\omega}'_j > \| < s$, we have a solution of $Ku_0 = 0$, such that $< u_0 \tilde{\omega}'_j > 0, j = 1, \dots, 2g$. Because of (19.16) such a solution is single-valued. As long as it has zero at q_0, $u_0(q) \equiv 0$.

From (20.1) it follows that $s \leqslant 2g$.

THEOREM 20.3. *A generalized analytic differential of the first kind is a solution of the equation $Kv = Z'$, where $Z'dz$ is an Abelian differential of the first kind. Conversely, any solution of the equation $Kv = Z'$ is a generalized analytic differential having a pole at the point p_0, with the principal part*

$$\frac{1}{\pi} \iint_{\widehat{M}} [A(p)v(p) + \overline{B(p)v(p)}] d\sigma_p = \frac{1}{\pi} < 1, v > - \frac{i}{\pi} < i, v > . \tag{20.2}$$

The first statement was proved in §19. The second one is analogous to (4.26).

Let $v_1, \dots, v_{s+h}$ be a real basis of the space $H(-p_0)$ of generalized analytic differentials which are multiples of the divisor $-p_0$, i.e., with the first order pole at p_0 and

$$\begin{aligned} \widetilde{K} v_j &= 0 \quad j = 1, \dots, s, \\ \widetilde{K} v_j &= Z'_j \, j = s+1, \dots, s+h. \end{aligned} \tag{20.3}$$

Differentials $dZ_j (j = s+1, \dots, s+h)$ form a real basis of the space L_a of Abelian differentials of the first kind satisfying the condition (19.27)

$$< u_j, Z' > 0, \qquad Ku_j = 0, \qquad j = 1, \dots, s. \tag{20.4}$$

THEOREM 20.4. *The real dimension of the space $H(-p_0)$ is equal to*

$$\dim H(-p_0) = s + h = 2g. \tag{20.5}$$

Indeed, the dimension h of the space L_a is equal to $2g - s$, where s is the rank of the matrix (20.1) (Theorem 20.2). This means that $\dim H(-p_0) = (2g - s) + s = 2g$.

Consider the matrix

$$\Sigma_0 = \begin{pmatrix} < 1, v_1 > & \dots & < 1, v_s > & \dots & < 1, v_{2g} > \\ < i, v_1 > & \dots & < i, v_s > & \dots & < i, v_{2g} > \end{pmatrix}. \tag{20.6}$$

THEOREM 20.5. *The real dimension l_0 of the space L_0 of generalized constants is equal to*

$$l_0 = 2 - \operatorname{rank} \Sigma_0. \tag{20.7}$$

Indeed, every zero linear combination of the lines of the matrix (20.6) corresponds to the constant c, such that

$$< c, v_j > 0, \qquad j = 1, \ldots, 2g. \tag{20.8}$$

The first s relations (20.8) involve the solvability of the equation $Ku_0 = c$. Further

$$< c, v_j >=< Ku_0, v_j >=< u_0, \widetilde{K} v_j > =< u_0, Z'_j >= 0, \qquad j = s+1, \ldots, 2g. \tag{20.9}$$

Equation (20.9) involves the existence of a single-valued solution of the equation $Ku = c$. Indeed, the general solution of this equation has the form

$$u = u_0 - \sum_{k=1}^{s} a_k u_k.$$

From (19.16) the single-valuedness conditions are

$$\begin{aligned} &\sum_{k=1}^{s} a_k < u_k, w'_\mu > =< u_0, w'_\mu >, \qquad \mu = 1, \ldots g, \\ &\sum_{k=1}^{s} a_k < u_k, i w'_\mu > =< u_0, i w'_\mu > . \end{aligned} \tag{20.10}$$

Let some combination of the lines of the matrix of the system (20.10) be equal to zero,

$$< u_1, Z' >= 0, \ldots, \qquad < u_s, Z' > 0.$$

This involves the solvability of the equation $Kv = Z'$ and hence $Z' \in L_a$ and $Z' = \sum_k c_k Z'_k$. Consequently, from (20.9), we have $< u_0, Z' >= 0$. This means that the system (20.10) is solvable. Hence every zero combination of the lines of (20.6) corresponds to an element of the space L_0. Conversely, every element of the space L_0 corresponds to some zero combination of the lines of (20.6).

THEOREM 20.6. *The real dimension h_0 of the space H_0 of generalized analytic differentials of the first kind is equal to*

$$h_0 = 2g - \operatorname{rank} \Sigma_0. \tag{20.11}$$

This statement follows from Theorem 10.4 and Eq. (20.2).

Theorem 20.6 and Theorem 20.5 imply the following theorem.

THEOREM 20.7.

$$l_0 - h_0 = 2 - 2g. \tag{20.12}$$

Therefore, the index of the Carleman–Vers–Vekua operator (19.2) coincides with the index of the Cauchy–Riemann operator $\bar{\partial}$. These results are natural since these operators are homotopic.

B. *The Riemann–Roch Theorem*

Let $D = \sum_{k=1}^{n} q_k$ be a positive divisor without multiple points (all points q_k are different), let $L(D)$ be the space of generalized analytic functions which are multiples of the divisor $-D$, and $H(D)$ be the space of generalized analytic differentials which are multiples of D.

Let $t_{q_k}(q)$ be the Abelian integral of the second kind (2.14).

THEOREM 20.8. *In order that the function $u \in L(D)$, it is necessary and sufficient that u be a single-valued solution of the equation*

$$Ku = t_D(q), \qquad t_D(q) = c_0 + \sum_{k=1}^{n} c_k t_{q_k}(q), \tag{20.13}$$

where $c_0, \ldots, c_n$ are arbitrary complex constants.

This statement is evident. The analog of the matrix Σ_0 for our case has the form

(20.14)

$$\sum = \begin{pmatrix} <1, v_1> & \ldots & <1, v_s> & \ldots & 0 & \ldots \\ <i, v_1> & \ldots & <i, v_s> & \ldots & 0 & \ldots \\ <t_{q_1} v_1> & \ldots & <t_{q_1}, v_s> & \ldots & <t_{q_1}, v_j> + \operatorname{Re} \pi Z_j'(q_1) & \ldots \\ \ldots & \ldots & \ldots & \ldots & \ldots & \ldots \\ <it_{q_n}, v_1> & \ldots & <it_{q_n}, v_s> & \ldots & <it_{q_n}, v_j> + \operatorname{Re} \pi i Z_j'(q_n) & \ldots \end{pmatrix}$$

The matrix has $2n+2$ lines and $2g$ columns.

THEOREM 20.9. *The real dimension of the space $L(D)$ is equal to*

$$\dim L(D) = 2n + 2 - \operatorname{rank} \sum. \tag{20.15}$$

PROOF: Every zero linear combination of the lines of (20.14) corresponds to the Abelian integral

$$t_D(q) = c_0 + \sum_{k=1}^{n} c_k t_{q_k}(q), \tag{20.16}$$

such that

$$<t_D, v_j>= 0, \qquad j = 1, \ldots, s, \tag{20.17}$$

$$< t_D, v_j > - \operatorname{Re} \pi \sum_{k=1}^{n} c_k Z_j'(q_k) = 0, \qquad j = s+1, \ldots, 2g. \tag{20.18}$$

From (20.17) the equation $Ku_0 = t_D$ is solvable. We obtain from Eq. (20.18)

(20.19)

$$\begin{aligned} &< t_D, v_j > + \operatorname{Re} \pi \sum_{k=1}^{n} c_k Z_j'(q_k) \\ &= < Ku_0, v_j > + Re\pi \sum_{k=1}^{n} c_k Z_j'(q_k) \\ &=< u_0, \widetilde{K} v_j > + \operatorname{Re} \pi \sum_{k=1}^{n} c_k Z_j'(q_k) \\ &= < u_0, Z_j' > + \operatorname{Re} \pi \sum_{k=1}^{n} c_k Z_j'(q_k) = 0, \qquad j = s+1, \ldots, 2g. \end{aligned}$$

The equation $Ku = t_D$ has a single-valued solution. In fact, the general solution of this equation has the form

$$u = u_0 - \sum_{n=1}^{s} \alpha_j u_j.$$

The conditions of the single-valuedness (19.16) lead to the system of equations

$$\begin{aligned} -\sum_{j=1}^{s} \alpha_j < u_j, iw_\mu' > &=< u_0, iw_\mu' > - \frac{1}{2} \operatorname{Re} \int_{k_{2\mu}} dt_D(q), \\ \sum_{j=1}^{s} \alpha_j < u_j, w_\mu' > &= - < u_0, w_\mu' > - \frac{1}{2} \operatorname{Re} i \int_{k_{2\mu}} dt_D(q). \end{aligned}$$

From (3.14) we obtain the system

$$\begin{aligned} -\sum_{j=1}^{s} \alpha_j < u_j, iw_\mu' > &=< u_0, iw_\mu' > + \operatorname{Re} \pi i \sum_{k=1}^{n} c_k w_\mu'(q_k), \\ -\sum_{j=1}^{s} \alpha_j < u_j, w_\mu' > &=< u_0, w_\mu' > + \operatorname{Re} \pi \sum_{k=1}^{n} c_k w_\mu'(q_k). \end{aligned} \tag{20.20}$$

Let some combination of the strings of the matrix of the system (20.20) be zero,

$$< u_j, Z' > 0, \qquad j = 1, \ldots, s.$$

This means that the equation $Kv_0 = Z'$ is solvable. From (20.19) the corresponding element of the augmented matrix is

$$< u_0, Z_j' > + \operatorname{Re} \pi \sum_{k=1}^{n} c_k Z_j'(q_k) = 0, \qquad j = s+1, \ldots 2g.$$

Hence the system (20.20) is solvable.

Conversely, let $uL(D), Ku = t_D$. The single-valuedness of this function means that

$$< u, iw'_\mu > - \frac{1}{2}\operatorname{Re}\int_{k_{2\mu}} dt_D = 0,$$
$$< u, w'_\mu > + \frac{1}{2}\operatorname{Re}\int_{k_{2\mu}} dt_D = 0, \qquad \mu = 1, \dots, g.$$

This involves the relations

$$(20.21)\qquad \begin{aligned} &< u, iw'_\mu > + \operatorname{Re}\pi i \sum_{k=1}^{n} c_k w'_\mu(q_k) = 0, \\ &< u, w'_\mu > + \operatorname{Re}\pi \sum_{k=1}^{n} c_k w'_\mu(q_k) = 0, \qquad \mu = 1, \dots, g. \end{aligned}$$

For Abelian differentials $dZ_j (j = s+1, \dots, 2g)$, we obtain

$$< u, Z'_j > + \operatorname{Re}\pi \sum_{k=1}^{n} c_k Z'_j(q_k) = 0,$$
$$< u, iZ'_j > + \operatorname{Re}\pi i \sum_{k=1}^{n} c_k Z'_j(q_k) = 0,$$

whence it follows that

$$< t_D, v_j > + \operatorname{Re}\pi \sum_{k=1}^{n} c_k Z'_j(q_k) = 0,$$
$$< it_D, v_j > + \operatorname{Re}\pi i \sum_{k=1}^{n} c_k Z'_j(q_k) = 0.$$

Therefore, any $u \in L(D)$ corresponds to zero combination of the lines of the matrix Σ.

THEOREM 20.10. *The real dimension of the space $H(D)$ is equal to*

$$(20.22)\qquad \dim H(D) = 2g - \operatorname{rank}\Sigma.$$

Every zero combination of the columns $v = \sum_{j=1}^{2g} \alpha_j v_j$ of Σ gives

$$(20.23)\qquad < 1, v >= 0, \qquad < i, v >= 0,$$

$$(20.24)\qquad \begin{aligned} &< t_{q_k}, v > + \operatorname{Re}\pi Z'(q_k) = 0, \qquad Z' = \sum_j \alpha_j Z'_j, \\ &< it_{q_k}, v > + \operatorname{Re}\pi i Z'(q_k) = 0, \qquad k = 1, \dots, n. \end{aligned}$$

Equation (20.21) means that $vdz(q)$ is regular at p_0 (see (20.2)). Taking into account that $M(q_k, p) = -t_{q_k}(p)$ (see (4.21)), we obtain from (20.24) that $v(q_k) = 0, k = 1, \dots, n$.

THEOREM 20.11 (RIEMANN–ROCH).

$$\dim L(D) - \dim H(D) = 2\deg D - 2g + 2. \tag{20.25}$$

For divisor $D = \sum_k q_k$ the statement follows from (20.22) and (20.15).

Let now D be an arbitrary divisor, $\deg D = n$. Choose a divisor $\widehat{D} = \sum_k q_k$. Assume that $D - \widehat{D} = \sum_k \alpha_k p_k, \deg(D - \widehat{D}) = 0$. Choose coefficient $A_0(p)$ satisfying the Eq. (19.13). Then the equation

$$\bar{\partial} u = A_0 u \tag{20.26}$$

has the solution u_0 whose divisor is $(u_0) = D - \widehat{D}$.

In this case the space $L(D)$ is transformed into the space $L(\widehat{D})$ for the equation

$$\bar{\partial} u = (A + A_0)u + B\frac{\bar{u}_0}{\bar{u}_0}u. \tag{20.27}$$

The space $H(D)$ is transformed into the space $H(\widehat{D})$ for the equation

$$\bar{\partial} v + (A + A_0)v + \bar{B}\frac{\bar{u}_0}{u_0}\bar{v} = 0. \tag{20.28}$$

For the spaces $L(\widehat{D})$ and $H(\widehat{D})$ the relation (20.25) has been proved.

Consider the Riemann boundary problems

$$u^+(p) = G(p)u^-(p), \tag{20.29}$$

$$u^+(p) = G(p)u^-(p) + g(p), \tag{20.30}$$

$$v^+(p)dz(p) = G^{-1}(p)v^-(p)dz(p), \qquad \operatorname{ind}_L G = \kappa, \tag{20.31}$$

for the equations (19.2) and (19.11), respectively.

THEOREM 20.12. *The number l of solutions of the problem (20.29) and the number h of solutions of the problem (20.31) satisfy the relation*

$$l - h = 2\kappa - 2g + 2. \tag{20.32}$$

For solvability of the problem (20.30) it is necessary and sufficient that

$$\operatorname{Re}\int_L g(p)v_j^+(p)dz(p) = 0, \qquad j = 1, \dots h, \tag{20.33}$$

where $v_1^+, \dots, v_h^+$ are the complete system of solutions of (20.31).

Let $f^{\pm}(q)$ be an analytic solution of the problem (20.29) (in general, having $g-\kappa$ poles). Denote the divisor $(f^{\pm}(q))$ by D. Then

$$\tilde{u}(q) = u(q)f^{-1}(q), \qquad \tilde{v}(q) = v(q)f(q) \tag{20.34}$$

are generalized analytic functions (differentials) satisfying the equations

$$\bar{\partial}\tilde{u} = A\tilde{u} + B\frac{\bar{f}}{f}\bar{\tilde{u}}, \tag{20.35}$$

$$\bar{\partial}\tilde{v} + A\tilde{v} + \bar{B}\frac{f}{\bar{f}}\bar{\tilde{v}} = 0. \tag{20.36}$$

The spaces of solutions of the problems (20.29) and (20.31) correspond to the spaces $L(D)$ and $H(D)$ for Eqs. (20.35) and (20.36), respectively. The problem (20.30) is reduced to the form

$$\tilde{u}^{+}(p) - \tilde{u}^{-}(p) = \frac{g(p)}{f^{+}(p)}, \qquad (\tilde{u}) + D \geqslant 0. \tag{20.37}$$

The necessity of the conditions (20.33) is verified directly. The sufficiency of (20.33), see [30], also Rodin [h],[i], and Koppelman [c].

§21 Nonlinear Aspects of the Generalized Analytic Function Theory

A. *Multiplicative Multivalued Solutions. Existence*

Let $u(s)$ be a generalized solution of the equation

$$\bar{\partial}u = Au + B\bar{u}, \qquad A, B \in L_q^{0,1}(M), \qquad q > 2. \tag{21.1}$$

It satisfies the equation

$$u(s) = \varphi(s)\exp -\frac{1}{\pi}\iint\limits_{\widehat{M}} \left[A(p) + B(p)\frac{\overline{u(p)}}{u(p)}\right] M_*(p,s)\,d\sigma_p, \tag{21.2}$$

where $M_*(p,s)$ is a multivalued kernel (4.4) and $\varphi(s)$ is analytic on $\widehat{M}$. As long as the kernel of this representation is multivalued, the function $\varphi(s)$ has, in general, multiplicative periods along the cycles $k_j (j = 1, \ldots, 2g)$. Conversely, substituting the value $\exp w(s)$ instead of $\varphi(s)$ ($w(s)$ is an Abelian integral of the first or third kind), we consider the equation (21.2) as a nonlinear integral equation and obtain a multiplicative multivalued solution of Eq. (21.1) on M. Fix the branch of the kernel by the condition (4.9). Then we get the solution corresponding to the chosen branch of the kernel. This branch satisfies Eq. (21.1). If all periods of the solution are equal to unity, we have a single-valued solution of the equation (21.1).

In particular, if $\varphi = \text{const}$, the equation

$$u(s) = c\exp -\frac{1}{\pi}\iint\limits_{\widehat{M}} \left[A(p) + B(p)\frac{\overline{u(p)}}{u(p)}\right] M_*(p,q)d\sigma_p \tag{21.3}$$

defines the solution having no zeros nor poles. We call such solutions *multiplicative constants.*

As opposed to the generalized constants, the multiplicative constants always exist.

Indeed, the operator

$$Ru = T(A + B\frac{\bar{u}}{u}) = -\frac{1}{\pi}\iint\limits_{\widehat{M}}\int \left[A(p) + B(p)\frac{\overline{u(p)}}{u(p)}\right] M_*(p,q)d\sigma_p \tag{21.4}$$

is compact in the space $L_2(M), \frac{1}{2} \leqslant \frac{1}{r} + \frac{1}{q} < 1$ and maps this space into itself. We get the estimate

$$\|Ru\|_{L_r} \leqslant M_{r,q}\|A + B\frac{\bar{u}}{u}\|_{L_q} \leqslant M_{r,q}\{\|A\|_{L_q} + \|B\|_{L_q}\} \tag{21.5}$$

for $\frac{1}{2} - \frac{1}{q} \leqslant \frac{1}{r} < 1$. This estimate means that the sphere of the radius

$$\rho > M_{r,q}\{\|A\|_{L_q} + \|B\|_{L_q}\}$$

is mapped into itself by the operator R. By the Schauder principle, the operator $c\exp Ru$ has a fixed point. As was shown above, this point is a multiplicative constant.

Therefore, we have the following theorem.

THEOREM 21.1. *The equation (21.1) is solvable for every constant in the space* $L(M), 1/2 \leqslant 1/q + 1/r < 1$. *The solution is the multiplicative constant satisfying the condition* $u(q_0) = c$.

B. *Multiplicative Constants. Uniqueness*

THEOREM 21.2. *The fixed point of the operator* $\exp R$ *is single in every space* $L_r(M), \frac{1}{2} \leqslant \frac{1}{q} + \frac{1}{r} < 1$.

Let $h(q) = -2\arg u(q)$. Then the equation $u = \exp Ru$ can be rewritten in the form

$$\begin{aligned} h(s) = Kh &= \iint\limits_{\widehat{M}} \{a(p,s) - b(p,s)\sin[h(p) + c(p,q)]\}d\sigma_p, \\ a(p,s) &= -\frac{2}{\pi}\operatorname{Im}[A(p)M_*(p,s)], \\ b(p,s) &= \frac{2}{\pi}|B(p)M_*(p,s)|, \\ c(p,s) &= \arg[B(p)M_*(p,s)]. \end{aligned} \tag{21.6}$$

If $h(s)$ is a solution of (21.6), then $u(q)$ is calculated by Eq. (21.3).

Let $h_0 = -2\arg u_0(q)$ be a fixed point of the operator K. The Frechet derivation of the operator K at the point h_0 is equal to

$$B\psi = -\iint_{\widehat{M}} b(p,s)\cos\left[h_0(p) + c(p,q)\right]\psi(p)\,d\sigma_p. \tag{21.7}$$

Now we study the operator B spectrum. We shall introduce the function Ψ,

$$\begin{aligned}&\operatorname{Im}\Psi(q) = \psi,\\ &\operatorname{Re}\Psi = \operatorname{Re}\frac{2i}{\pi}\iint_{\widehat{M}} B_0(p)\psi(p)M_*(p,q)\,d\sigma_p.\end{aligned}$$

Then the equation $B\psi = \lambda\psi$ can be rewritten in the form

$$\begin{aligned}&\Psi(q) - \frac{1}{\pi}\iint_{\widehat{M}} B_0(p)\left[\Psi(p) - \overline{\Psi(p)}\right]M_*(p,q)\,d\sigma_p = 0,\\ &B_0(p) = -\frac{2}{\lambda}B(p)\frac{\overline{u_0(p)}}{u_0(p)}.\end{aligned} \tag{21.8}$$

LEMMA. *The equation (21.8) does not have any nontrivial solutions for* $B_0 \in L_q^{0,1}, q > 2$.

We shall show that the adjoint equation

$$v(q) - \frac{1}{\pi}\iint_{\widehat{M}} \left[B_0(p)v(p) - \overline{B_0(p)v(p)}\right]M_*(q,p)\,d\sigma_p = 0 \tag{21.9}$$

has no nontrivial solutions. If $v(q)$ is a solution of Eq. (21.9), then

$$\bar{\partial}v - B_0 v + \overline{B_0 v} = 0. \tag{21.10}$$

In this case the form $iw = v\,dz - \overline{v\,dz}$ is closed, i.e., $w = df$. This means that $v = -i\partial f$ and $\bar{v} = i\bar{\partial}f$. Therefore, the value $f(q)$ is real and satisfies the equation

$$\frac{\partial^2 f}{\partial z(q)\partial\overline{z(q)}} + B_0\frac{\partial f}{\partial z(q)} + B_0\frac{\partial f}{\partial\overline{z(q)}} = 0. \tag{21.11}$$

As long as $v(q)$ may possess a single singularity at the point $p = s_0$, the value $f(q)$ has a finite upper or lower bound on M. Rewrite (21.9) in the form

$$i\partial_q f(q) - \frac{1}{\pi}\iint_{\widehat{M}} \bar{\partial}_p v(p)\partial_q\left[\Omega_{s_0 q}(p) - \Omega_{s_0 q}(q_0)\right]d\sigma_p = 0, \tag{21.12}$$

and suppose

$$(21.13) \qquad iF(q) = if(q) - \frac{1}{\pi}\iint\limits_{\widehat{M}} \bar{\partial}_p v\,[\Omega_{s_0p}(p) - \Omega_{s_0q}(q_0)]\,d\sigma_p.$$

The function $F(q)$ is analytic on M. The second addendum of the right-hand side of (21.13) is single-valued on M. Indeed, from (4.2), (2.13′), and (3.12) we have

$$\int\limits_{k_j} d_q[\Omega_{s_0q}(p) - \Omega_{s_0q}(q_0)] = \int\limits_{k_j} d\Omega_{q_0p}(q) - 2\pi i \sum_{l=1}^{g}\Big\{\operatorname{Im}\int_{q_0}^{p} d\theta_{2l-1}\operatorname{Im}\int\limits_{k_j} d\theta_{2l}$$
$$- \operatorname{Im}\int_{q_0}^{p} d\theta_{2l}\operatorname{Im}\int\limits_{k_j} d\theta_{2l-1}\Big\} = 0, \qquad j = 1,\ldots,g.$$

Therefore, periods of $F(q)$ coincide with the periods of $f(q)$, and hence they are real. This means that $F =$ const. Whence it follows that the function $f(q)$ is single-valued. By the maximum principle for the equation (21.11) (see Bers & Nirenberg [a], Bers, John, & Schechter [4]), $f \equiv$ const, and hence $v \equiv 0$.

Show briefly how one may get the theorem statement (for details, see Krasnoselskii [22], chapter 2, 3).

The point h_0 is an isolated fixed point of the operator K. Indeed, let S_ε be a sphere with the radius ε with the centre at h_0 in $L_r(M)$, $\frac{1}{2} \leqslant \frac{1}{r} + \frac{1}{q} < 1$ and $h_0 + g$ be a point of the sphere. Then

$$\begin{aligned}\|K(h_0+g) - (h_0+g)\| &= \|K(h_0+g) - Kh_0 - g\| \\ &= \|Bg - g + \alpha g\| \geqslant \|Bg - g\| - \|\alpha g\| > \varepsilon_1\|g\|,\end{aligned}$$

since 1 is no eigenvalue of the operator B. This means that the operator K has no fixed points on S_ε for small ε. The operator $E - K$ is homotopic to $E - B$ on S_ε, since the vector field

$$\Phi(g,t) = t[(h_0+g) - K(h_0+g)] + (1-t)[(h_0+g) - (h_0+Bg)]$$

has no zeros on S_ε and $\Phi(g,0) = g - Bg$ and $\Phi(g,1) = (h_0+g) - K(h_0+g)$. Therefore, the rotations of the operators K and B coincide on S_ε. The rotation of K on S_ε is equal to the index of the point h_0; the rotation of B on S_ε is equal to $(-)^\beta$, where β is the sum of the multiplicities of all eigenvalues of the operator B belonging to the set $[1,\infty)$. As has been shown above, $\beta = 0$, and hence the index of any fixed point of the operator K is equal to $+1$. As long as the operator K is compact, its fixed points form a compact set. Since this set is discrete, it is finite. Therefore, the operator K has a finite number of fixed points. As long as the operator K maps the sphere S_R of a large radius R into itself, this operator is homotopic to the unit operator E on S_R. Hence, the rotation of K on S_R is equal to $+1$. On the other hand, the rotation K on S_R is equal to the sum of the fixed point indices. Whence it follows that the fixed point of the operator K is the unique one.

C. *Abel's Theorem*

Abel's problem of the existence of the solution of the equation (23.1), determined by the divisor $\gamma = \sum_1^n \alpha_k p_k, \deg\gamma = 0$, is reduced to the integral equation

$$u(q) = \exp\left\{\sum_{k=1}^{n} \alpha_k \Omega_{p_0 p_k}(q) \quad - \sum_{j=1}^{g} c_j w_j(q)\right\} Ru, \tag{21.15}$$

where $w_j(q), j = 1, \ldots, g$ is a basis of Abelian integrals of the first kind and $\Omega_{p_0 p_k}$ are Abelian integrals of the third kind. Let

$$v(q) = u(q)H(q), H(q) = \exp\left\{\sum_{k=1}^{n} \alpha_k \Omega_{p_0 p_k}(q) + \sum_{j=1}^{g} c_j w_j(q)\right\}. \tag{21.16}$$

We get the equation $v = \widetilde{R}v$. The operator $\widetilde{R}$ may be obtained from the operator R by the exchange of the coefficient $B(q)$ by $B(q)\overline{H(q)}/H(q)$. Let $v(c_1, \ldots c_g)$ be a fixed point of the operator R with multiplicative periods $v_j(c_1, \ldots, c_g)$ along the cycles $k_j (j = 1, \ldots, 2g)$. Abel's problem is reduced to the system of equations

$$v_j(c_1, \ldots, c_g) = 1, \qquad j = 1, \ldots, g. \tag{21.17}$$

CHAPTER 6

INTEGRABLE SYSTEMS

§22 Schrödinger Equation

A. *Fast-Decreasing Potentials*

Consider the one-dimensional Schrödinger equation

$$L(t)\psi \equiv -\frac{d^2\psi}{dx^2} + u(x,t)\psi = \lambda\psi, \qquad \lambda = k^2 \tag{22.1}$$

with a real potential $u(x,t)$ satisfying the condition

$$\int_{-\infty}^{\infty} (1+|x|)|u(x,t)|dx < \infty. \tag{22.2}$$

Such potentials are called *fast-decreasing.*

We shall consider the operator family $L(t)$ whose spectrum is independent of t. For this it is sufficient that there exists a family of unitary operators $\{U(t)\}$, such that the operators $L(t)$ be a unitarily equivalent,

$$L(t) = U^*(t)L(0)U(t). \tag{22.3}$$

It is clear that all spectral characteristics coincide in this case. In order that the family $\{U(t)\}$ exists, it is necessary and sufficient that the operators $L(t)$ satisfy the Lax equation

$$\frac{dL}{dt} = LA - AL = [L, A], \tag{22.4}$$

where A is a skew-symmetric operator.

Indeed, as long as the operators $U(t)$ are unitary,

$$\dot{U}^* = (U^{-1})^{\cdot} = -U^{-1}\dot{U}U^{-1} = -U^*\dot{U}U^*,$$

and hence

$$\begin{aligned}\dot{L}(t) &= -U^*\dot{U}U^*L(0)U + U^*L(0)\dot{U}\\ &= -U^*\dot{U}L(t) + L(t)U^*\dot{U} = LA - AL,\end{aligned}$$

where the operator $A = U^*\dot{U}$ is skew-symmetric,

$$A^* = \dot{U}^*U = -U^*\dot{U}U^*U = -A.$$

Conversely, if $U(t)$ is a unitary solution of the equation

$$\dot{U} = UA, \qquad U(0) = 1,$$

and $L(t)$ satisfies (22.4), Eq. (22.3) is valid, since $A = U^*\dot{U}$.

As follows from (22.1), the operator $\dot{L}$ is the operator of multiplication by the function $\dot{u}(x,t)$. Hence the right side [L,A] of the Lax equation (22.4) has to be a multiplication by a function operator. We suppose that A is a differential operator of order $2q+1$ (the operator order is odd, since the operator is skew-symmetric). If $q = 0$, $A = \frac{d}{dx}$, and $[A, L] = u'_x$. We get the equation

$$\dot{u} = u_x, \qquad u(x,t) = f(x+t).$$

The first nontrivial operator appears for $q = 1$. In this case we have

$$\begin{aligned}A &= 4\frac{d^3}{dx^3} - 6u\frac{d}{dx} - 3u'_x,\\ [A, L] &= 6uu_x - u_{xxx}.\end{aligned} \tag{22.5}$$

The Lax equation has the form

$$u_t = 6uu_x - u_{xxx}. \tag{22.6}$$

It is the famous Korteweg–de Vries (KdV) equation, known in many areas of nonlinear hydrodynamics [24], [40]. In particular, the separate waves in canals based on the theory of solitons were investigated by Scott Rassel.

For any $q > 1$, there exists the operator

$$A_q = \frac{d^{2q+1}}{dx^{2q+1}} + \sum_{j=1}^{\infty}\left[b_j\frac{d^{2j-1}}{dx^{2j-1}} + \frac{d^{2j-1}}{dx^{2j-1}}b_j\right], \tag{22.7}$$

such that $[A_q, L]$ is a multiplication by the function operator, and b_j are polynomials with respect to $u(x,t)$ and its derivatives of x. The equations

$$\dot{L} = [A_q, L], \qquad q > 1$$

are called the higher KdV equations. For simplicity, we restrict ourselves to the cases (22.5) and (22.6).

We consider the Cauchy problem for Eq. (22.6) and suppose that the initial value $u(x,0)$ satisfies the condition (22.2). For any fixed t, the problem of the calculation of $u(x,t)$ is equivalent to the inverse scattering problem for Eq. (22.1). This was investigated by Gardner, Green, Kruskal & Miura [a]. In turn, the inverse scattering problem is reduced to the matrix Riemann problem whose coefficients dynamics is very simple. Solving this problem for every x, t, we get the solution of the nonlinear KdV equation.

The operators $L(t)$ are self-adjoint, their discrete spectrum is finite and negative (as is generally known, the eigenvalues of the Schrödinger operator correspond to bounded states of the particle), and the continuous spectrum described by the scattering theory is the positive semi-axis $\lambda \geqslant 0$(or the real axis $\operatorname{Im} k = 0$).

The Jost functions for the Eq. (22.1) are defined by the asymptotic behaviour

$$f_j(x,k) = \exp(-ikx) + o(1) \qquad \text{as} \quad x \longrightarrow (-1)^j\infty. \tag{22.8}$$

For $\lambda \geqslant 0$ the pairs of the functions $(f_1, \bar{f}_1), (f_2, \bar{f}_2)$ form complete systems of the solutions of the equation (22.1). Therefore, f_1 is represented in the form

$$f_1(x,k) = a(k) f_2(x,k) + b(k)\overline{f_2(x,k)}. \tag{22.9}$$

This relation is easily interpreted physically. Rewrite this equation in the form

$$\frac{f_1(x,k)}{a(k)} = f_2(x,k) + \frac{b(k)}{a(k)}\overline{f_2(x,k)}.$$

The wave $\exp(-ikx)$ moving from right to left is scattered by the potential $u(x,t)$ and generates two waves: the transmitted wave with the asymptotic behaviour

$$a^{-1}\exp(-ikx) \qquad \text{at} - \infty,$$

and the reflected wave with the asyptotic behaviour

$$b(k)a^{-1}(k)\exp(ikx) \qquad \text{at} + \infty.$$

So the value $a^{-1}(k)$ is called the *transmitted coefficient*, $b(k)a^{-1}(k)$ is called the *reflection coefficient*, and $a(k)$ is the *scattering amplitude*. There are many books that treat the scattering theory in detail [6], [11], and [40].

Note that if f and g are two linearly independent solutions of (22.1), the Wronskian $W(f,g)$ is independent of x. Therefore, $W(f_j, \bar{f}_j) = 2ik, j = 1, 2$ (calculate this value at $(-1)^j\infty$). This means that the determinant of the matrix

$$T(k) = \begin{pmatrix} a(k) & \overline{b(k)} \\ b(k) & \overline{a(k)} \end{pmatrix}$$

is equal to unity,

$$|a(k)|^2 - |b(k)|^2 = 1, \qquad \operatorname{Im} k = 0.$$

Consider the functions

$$\psi_j(x,k) = f_j(x,k)\exp ikx, \qquad j = 1,2. \tag{22.10}$$

The function $\psi_1(x,k)$ is analytic in the upper semiplane $\operatorname{Im} k > 0$, and the function $\psi_2(x,k)$ is analytic in the lower semiplane $\operatorname{Im} k < 0$. As long as the potential $u(x,t)$ is unessential for $\lambda \longrightarrow \infty$, $f_j(x,k) \sim \exp(-ikx)$, and hence $\psi_j(x,\infty) = 1, j = 1,2$.

The representation

$$\begin{gathered} f_j(x,k) = e^{-ikx} + \frac{(-1)^j}{k}\int_x^{\infty_j} \sin k(x-\xi)u(\xi,t)f_j(\xi,k)d\xi, \\ \infty_j = (-1)^j\infty \end{gathered} \tag{22.11}$$

can be verified directly. We get the equation

$$\psi_j(x,k) = 1 + \frac{(-1)^j}{2ik}\int_x^{\infty_j}[e^{2ik(x-\xi)} - 1]\psi_j(\xi,k)u(\xi,t)d\xi. \tag{22.12}$$

The representation (22.12) allows to continue analytically the functions $\psi_j(x,t)$ into the upper (for $j = 1$) and lower (for $j = 2$) halfplanes. As long as (see (22.9))

$$a(k) = \frac{1}{2ik}\begin{vmatrix} f_1(x,k) & \bar{f}_2(x,k) \\ f_{1x}(x,k) & \bar{f}_{2x}(x,k), \end{vmatrix} \tag{22.13}$$

$$b(k) = \frac{1}{2ik}\begin{vmatrix} f_2(x,k) & f_1(x,k) \\ f_{2x}(x,k) & f_{1x}(x,k), \end{vmatrix} \tag{22.14}$$

the scattering amplitude $a(k)$ is analytic in the upper halfplane, and $a(\infty) = 1$. The zeros $k_s (s = 1,\ldots,n)$ of the scattering amplitude $a(k)$ corresponds to the eigenvalues of the operator L, since $a(k_s) = 0$ entails the relations

$$\begin{gathered} f_1(x,k_s) = \beta_s\bar{f}_2(x,k_s), \\ f_{1x}(x,k_s) = \beta_s\bar{f}_{2x}(x,k_s), \\ s = 1,\ldots,n. \end{gathered} \tag{22.15}$$

This means that the function $f_1(x,k_s)$ has the asymptotic behaviour

$$\begin{aligned} f_1(x,k_s) &\sim \exp(-ik_s x) &&\text{as} \quad x \longrightarrow -\infty, \\ f_1(x,k_x) &\sim \beta_s \exp ik_s x &&\text{as} \quad x \longrightarrow +\infty, \end{aligned}$$

and is an eigenfunction of $L(t)$. As long as the eigenvalues $\lambda_s = k_s^2$ are negative, the zeros k_s are purely imaginary. It can be shown that all zeros of the scattering amplitude are simple.

Let us introduce vector lines that are analytic in the upper (lower) halfplane,

$$\begin{aligned} \psi_+(k) &= (\psi_1(x,k), \bar{\psi}_2(x,k)), \\ \psi_-(k) &= (\psi_2(x,k), \bar{\psi}_1(x,k)). \end{aligned} \tag{22.16}$$

These vectors are connected on the real axis $\operatorname{Im} k = 0$ by the relation

$$\begin{aligned} \psi_+(k) &= (f_1(x,k), \bar{f}_2(x,k)e^{ikx\sigma_3}) \\ &= \Big(\frac{1}{a(k)} f_2(x,k) + \frac{b(k)}{a(k)} \bar{f}_1(x,k), \\ &\qquad -\frac{\overline{b(k)}}{a(k)} f_2(x,k) + \frac{1}{a(k)} \bar{f}_1(x,k)\Big)e^{ikx\sigma_3} \\ &= (f_2(x,k), \bar{f}_1(x,k)) \frac{1}{a(k)} \begin{pmatrix} 1 & -\overline{b(k)} \\ b(k) & 1 \end{pmatrix} e^{ikx\sigma_3} \\ &= \psi_-(k)\frac{G(k)}{a(k)}, \qquad \sigma_3 = \begin{pmatrix} 1 & 0 \\ 0 & -1 \end{pmatrix}, \\ G(k) &= \exp(-ikx\sigma_3) \begin{pmatrix} 1 & -\overline{b(k)} \\ b(k) & 1 \end{pmatrix} \exp ikx\sigma_3. \end{aligned} \tag{22.17}$$

Therefore, we obtain the Riemann matrix problem on the complex plane k with the contour $\operatorname{Im} k = 0$. This problem may be also considered as the problem on the Riemann surface of $\lambda = k^2$ with the contour $\{\operatorname{Im}\lambda = 0, \operatorname{Re}\lambda \geqslant 0\}$.

Thus, the values $\{b(k), k_s, \beta_s, s = 1, \ldots, n\}$ allow us to calculate the potential $u(x,t)$. These values are called the *scattering data*.

In order to make the statement of the Riemann problem more exact, we use the operator

$$f^*(x,k) = \frac{df}{dx} + ikf(x,k), \tag{22.18}$$

and the matrices

$$\begin{aligned} \Psi_+(k) &= \begin{pmatrix} f_1(x,k) & \bar{f}_2(x,k) \\ f_1^*(x,k) & [\bar{f}_2(x,k)]^* \end{pmatrix} \exp ikx\sigma_3, \\ \Psi_-(k) &= \begin{pmatrix} f_2(x,k) & \bar{f}_1(x,k) \\ f_2^*(x,k) & [\bar{f}_1(x,k)]^* \end{pmatrix} \exp ikx\sigma_3. \end{aligned} \tag{22.19}$$

Evidently, the matrices $\Psi_\pm(k)$ satisfy the equation

$$\begin{aligned} (\Psi_\pm)_x &= -ik\sigma_3\Psi_\pm + ik\Psi_\pm\sigma_3 + Q(x,t)\Psi_\pm, \\ Q(x,t) &= \begin{pmatrix} 0 & 1 \\ u & 0 \end{pmatrix}, \end{aligned} \tag{22.20}$$

the boundary condition (22.17), and

$$\Psi_\pm(k) \sim \begin{pmatrix} 1 & 1 \\ 0 & 2ik \end{pmatrix} \qquad \text{as} \quad k \longrightarrow \infty. \tag{22.21}$$

The columns of $\Psi_+(k)$ are linearly dependent at the points $k_s (s = 1, \ldots, n)$ with coefficients β_s. We refer the reader to Chapters 1 and 3 of the monograph [40].

We shall now study the dynamics of the coefficients of the Riemann problem. Differentiating the equation (22.1), we get the relation

$$\dot{L}f + L\dot{f} = k^2 \dot{f}.$$

Taking into account Eq. (22.4), we have

$$(L - k^2)(\dot{f} + Af) = 0. \tag{22.22}$$

Therefore, if f is a solution of (22.1), then the function

$$\tilde{f} = \dot{f} + A_f \tag{22.23}$$

is a solution of Eq. (22.1). Denote

$$\tilde{f}_j = \dot{f}_j + Af_j, \qquad j = 1, 2.$$

If $x \longrightarrow -\infty$, $\dot{f}_1(x,k) \longrightarrow 0$, and hence

$$\tilde{f}_1(x,k) = 4ik^3 e^{-ikx} + o(1) \qquad \text{as} \quad x \longrightarrow -\infty.$$

Thus, $\tilde{f}_1 = 4ik^3 f$, and

$$\dot{f}_1(x,k) = 4ik^3 f_1(x,k) - Af_1(x,k).$$

Because of (22.9), as $x \longrightarrow +\infty$, we get the relation

$$\dot{a}(k)e^{-ikx} + \dot{b}(k)e^{ikx} = (4ik^3 - 4\frac{d^3}{dx^3})[a(k)e^{-ikx} + b(k)e^{ikx}]$$

that entails the Gardner–Green–Kruskal–Miura (GGKM) equations

$$\dot{a}(k) = 0, \qquad \dot{b}(k) = 8ik^3 b(k). \tag{22.24}$$

For $t = 0$, the values $a(k), b(k)$ are calculated by the initial value $u(x,0)$. Then one calculates $a(k), b(k)$ for any moment t and solves the Riemann problem by any approximation method. Finally, the potential is calculated by the formula

$$u(x,t) = f^{-1}(x,0)\frac{d^2 f(x,0)}{dx^2}. \tag{22.25}$$

Instead of the Riemann problem, one can solve the integral equation known as the Gelfand–Levitan–Marchenko equation [6], [40].

Consider in greater detail the case $b(k) = 0$ (reflectionless potential). Corresponding potentials are called solitons. The problem (22.17)–(22.21)–(22.15) is reduced to the scalar problem

$$\psi_1(x,k) = \frac{1}{a(k)}\psi_2(x,k), \qquad \psi_j(x,\infty) = 1.$$

Introduce the analytic function

$$R(k) = \begin{cases} \psi_1(x,k) & \operatorname{Im} k \geqslant 0 \\ [\bar{a}(k)]^{-1}\psi_2(x,k) & \operatorname{Im} k \leqslant 0. \end{cases} \tag{22.26}$$

The function $R(k)$ is analytic, has poles at $\bar{k}_1, \dots, \bar{k}_n$, and $R(\infty) = 1$. We confine ourselves to the single-soliton case $n = 1$. Then

$$R(k) = 1 + \frac{B(x,t)}{k - \bar{k}_1}, \tag{22.27}$$

and

$$\begin{aligned} f_1(x,k) &= (1 + \frac{B(x,t)}{k - k_1}) \exp(-ikx), \\ f_2(x,k) &= (1 + \frac{B(x,t)}{k - k_1}) \bar{a}(k) \exp(-ikx). \end{aligned} \tag{22.28}$$

The function $a(k)$ is analytic in the upper halfplane, has zero at $k + k_1$, and is real on the axis $\operatorname{Im} k = 0$. Therefore,

$$a(k) = \frac{k - k_1}{k - \bar{k}_1}. \tag{22.29}$$

By the relation (22.15)

$$(1 + \frac{B}{k - \bar{k}_1}) \exp(-ik_1 x) = \beta \frac{\bar{B}}{k - \bar{k}_1} \exp(ikx). \tag{22.30}$$

Solving the equation (22.30), we calculate $f_1(x,k)$. Further, using the relation (22.25), we get the one-soliton solution

$$\begin{aligned} u(x,t) &= -\frac{2\kappa^2}{ch^2\kappa(x - 4\kappa^2 t - \varphi}), \\ \varphi &= \frac{1}{2\kappa} \ln \frac{\beta}{2\kappa}. \end{aligned} \tag{22.31}$$

The function $u(x,t)$ is a nonlinear wave moving with the speed $4\kappa^2$ and with an amplitude of $2\kappa^2$, depending on the speed. The multisoliton solution has the following form at infinity,

$$u(x,t) = -\sum_{j=1}^{n} \frac{2\kappa_j}{ch^2\kappa_j(x - 4\kappa_j^2 - \varphi_j)}. \tag{22.31$'$}$$

Nonlinear addendums of (22.31′) interact nonellastically and conserve their form. They act as particles that give rise to the term "soliton."

B. *Reflection Finite-Zone Potentials*

In this section, instead of the condition (22.2), we assume that a potential $u(x,t)$ is bounded on the x-axis. This leads to hyperelliptic Riemann surfaces of a nonzero genus.

We shall base this on the finite-zones integration method investigated in the 1970s by Novikov and others (see [9],[40], Novikov [a], Dubrovin [a,b], Its & Matveev [a], McKean & Van Moerbeke [a], Novikov & Dubrovin [a,b], Krichever & Novikov [a], Lax [a,b], Marchenko [a,b], Marchenko & Ostrovskii [a], Meiman [a]. See also Khruslov [a], Firsova [a], Villalon [a], Cotlarov & Khruslov [a], Ermakova [a]). Let $\theta(x,\lambda)$ and $\varphi(x,\lambda)$ be the solutions of Eq. (22.1) satisfying the initial conditions

$$\theta(0,\lambda)=\varphi'_x(0,\lambda)=1, \qquad \theta'_x(0,l)=\varphi(0,\lambda)=1. \tag{22.32}$$

As is well known from [25], [26], and [33], for any nonreal λ, there exist the solutions

$$f_j(x,\lambda)=\theta(x,\lambda)+m_j(\lambda)\varphi(x,\lambda), \qquad j=1,2, \tag{22.33}$$

which are integrable over the semi-axes $(-\infty,0)$for $j=1$, and $(0,\infty)$ for $j=2$. Here m_j are the so-called Weyl–Titchmarsh functions. These functions are represented by

$$\begin{gathered} m_j(\lambda)=\int_{-\infty}^{\infty}\left[\frac{1}{\lambda-\tau}+\frac{\tau}{1+\tau^2}\right]d\rho_j(\tau)+a_j, \qquad j=1,2,\\ m_j(\lambda)=(-1)^j i\sqrt{\lambda}+\rho_j(-\infty)+o(1) \qquad \text{as} \quad \lambda\longrightarrow\infty. \end{gathered} \tag{22.34}$$

The functions $\rho_j(\tau)$ are the spectral functions of the operator $L(t)$ in the spaces $L_2(-\infty,0)$ for $j=1$, and $L_2(0,\infty)$ for $j=2$. Spectral functions are nondecreasing, absolutely continuous functions with finite gaps at the points of the discrete spectrum which are determined by the conditions

$$\begin{gathered} \int_{-\infty}^{\infty}\varphi(x,\lambda)\varphi(y,\lambda)d\rho_j(\lambda)=\delta(x-y),\\ x,y\in\begin{cases}(-\infty,0) & j=1,\\ (0,\infty) & j=2.\end{cases} \end{gathered} \tag{22.35}$$

A spectral matrix $(\xi_{ij}(\tau))$ of the operator $L(t)$ in the space $L_2(-\infty,\infty)$ is a symmetrical matrix with absolutely continuous elements possessing the finite gaps at the discrete spectrum points and determined by the conditions

$$\begin{aligned} \int_{-\infty}^{\infty}&[\theta(x,\lambda)\theta(y,\lambda)d\xi_{11}(\lambda)+[\varphi(x,\lambda)\theta(y,\lambda)\\ &+\theta(x,\lambda)\varphi(y,\lambda)]d\xi_{12}+\varphi(x,\lambda)\varphi(y,\lambda)d\xi_{22}(\lambda)]=\delta(x-y). \end{aligned} \tag{22.35'}$$

The Weyl–Titchmarsh functions may be represented in the form

$$m_j(\lambda) = \frac{(-1)^j \frac{1}{2} + M_{12}(\lambda)}{M_{11}(\lambda)}$$

(22.36)

$$M_{1j} = \int_{-\infty}^{\infty} \frac{d\xi_{1j}(\tau)}{\lambda - \tau}, \qquad j = 1, 2.$$

From Eqs. (22.34) and (22.36) it follows that the functions $m_j(\lambda)$ have poles on the real axis at the eigenvalues of the operator L in the spaces $L_2(-\infty, 0)$ and $L_2(0, \infty)$. The functions $m_j(\lambda)$ of the complex variable λ have a finite gap on the continuous spectrum E zones of the operator L in the space $L_2(-\infty, \infty)$. These functions may possess logarithmic singularities at the ends of the zones of E. Because of (22.4), the functions $m_j(\lambda)$ are time-independent.

We shall consider the Cauchy problem for the equation (22.4), on the assumption that the initial value of the potential $u(x, 0)$ is continuous and that the operator $L(0)$ (and hence $L(t)$) possesses the continuous spectrum

$$E = \bigcup_{\kappa=0}^{n-1} (\lambda_{2k}, \lambda_{2\kappa+1}) \bigcup (\lambda_{2n}, \infty),$$

consisting of a finite number of zones. The discrete spectra of the operator L in the spaces $L_2(-\infty, 0)$ and $L_2(0, \infty)$ are supposed to be finite. This class of potentials is called *reflection finite zones* and is denoted by R (Rodin [w], [30]).

The complete description of this class is still unknown, but it is known to contain some potential classes which have already been described.

First of all, the class R contains Novikov's finite-zone periodic and almost periodic potentials [9], [40]. In the periodic case, the Weyl–Titchmarsh functions have the form (cf. [25], [33]),

$$m_j(\lambda) = \frac{Q(\lambda)}{P(\lambda)} - (-1)^j i \frac{\sqrt{R(\lambda)}}{P(\lambda)}, \qquad j = 1, 2,$$

(22.37)

$$R(\lambda) = \prod_{l=0}^{2n} (\lambda - \lambda_l), \qquad P(\lambda) = \prod_{k=1}^{n} (\lambda - \xi_k),$$

$$Q(\lambda) = P(\lambda) \sum_{k=1}^{n} \frac{\sqrt{-R(\xi_k)}}{P'(\xi_k)(\lambda - \xi_k)}.$$

Here λ_k are the ends of the spectrum zones, and $\xi_k \in (\lambda_{2k-1}, \lambda_{2k})$ are arbitrary points belonging to lacunas. The Weyl solution (22.33) coincides with the Floke solutions, i.e., eigenfunctions of the operator of the shift on the period in the space of the solutions of Eq. (22.1).

Class R of reflection finite-zone potentials contains also other potentials. For example, the disturbance of a finite- zone periodic potential by a fast-decreasing potential leads to a finite-zone potential whose discrete spectrum belongs to lacunas (see, for example, Rofe-Beketov [a]). A disturbance by integrable potentials also preserves the finite-zone spectrum, but a discrete spectrum may belong both to lacunas and to zones. Note that the periodical potential on the semi-axis,

$$u(x) = \begin{cases} 0 & x < 0 \\ q(x) & x > 0, \end{cases} \qquad q(x) \text{ is periodical,}$$

may also have a finite-zones spectrum (see Gehtman & Stankevich [a]).

Let $u_1(x)$ and $u_2(x)$ be finite-zones periodic potentials with continuous spectra E_1 and E_2, respectively. There are two bounded nonnormalizable eigenfunctions for any point of these spectra. Then the operator L with potential

$$u(x) = \begin{cases} u_1(x) & x < 0 \\ u_2(x) & x \geqslant 0 \end{cases}$$

has the continuous spectrum $E = E_1 \cup E_2$. In the domain $E_1 \cap E_2$ there are also two bounded eigenfunctions. In the domain $E \backslash (E_1 \cap E_2)$, one eigenfunction is bounded and another is unbounded as $x \longrightarrow \infty$ (or $x \longrightarrow -\infty$). Moreover, it is possible that added eigenvalues appear in forbidden zones (analogs to Tamm states). The reader may verify these facts, assuming that u_1 and u_2 are constants.

Consider the two-sheeted Riemann surface M of the function

$$w^2 = R(\lambda), \qquad R(\lambda) = \prod_{j=0}^{2n} (\lambda - \lambda_j).$$

We shall use the local coordinate ς at infinity,

$$\varsigma = k^{-1}, \qquad k = \sqrt{\lambda - \lambda_{2n}}. \tag{22.38}$$

Sheets of M are determined by the sign of $\operatorname{Im} k$ for $|\lambda| > |\lambda_{2n}| + |\lambda_0|$. If $\operatorname{Im} k > 0$, we have the sheet M_1, and the sheet M_2 corresponds to $\operatorname{Im} k < 0$. The functions $f_j(x, \lambda)$ have the following asymptotes for large λ (the potential $u(x,t)$ is immaterial),

$$f_j(x, \lambda) = \exp(-ikx)\varphi_j(t) + o(1) \qquad \text{as} \quad \lambda \longrightarrow \infty, \qquad j = 1, 2. \tag{22.39}$$

Here $\varphi_j(t)$ is some function of time. As long as the functions $f_j(x, \lambda)$ have to be integrable with respect to x, we conclude that in this formula $\lambda \in M_j (j = 1, 2)$.

We introduce the functions

$$\bar{f}_j(x, \lambda) = \overline{f_j(x, \tilde{\lambda})}, \qquad \tilde{\lambda} \in M_j, \qquad j = 1, 2. \tag{22.40}$$

The operation $\sim$ consists of the passage from the point λ to $\bar{\lambda}$ and the following projection on the other sheet of the surface.

We have the relation

$$f_1(x, \lambda_+) = a(\lambda) f_2(x, \lambda_-) + b(\lambda) \overline{f_2(x, \lambda_-)} \tag{22.41}$$

on the cut E. Here $\lambda_\pm$ are boundary points of the sheets M_1 (the sign +) and M_2 (the sign −). By the gluing rule, if the point λ_+ belongs to the upper bar of the cut E on M_1, the corresponding point λ_- belongs to the lower bar on M_2, and conversely. As follows from (22.32),

$$\begin{aligned} a(\lambda) + b(\lambda) &= 1, \\ a(\lambda) m_2(\lambda_-) + b(\lambda)\overline{m_2(\lambda_-)} &= m_1(\lambda_+). \end{aligned} \tag{22.42}$$

The relation (22.41) means that the pairs of linear independent solutions

$$(f_1(x,\lambda_+), \overline{f_1(x,\lambda_+)}), \qquad (f_2(x,\lambda_-), \overline{f_2(x,\lambda_-)})$$

are related on E by the transition matrix

$$\begin{aligned} (f_1, \bar f_1) &= (f_2, \bar f_2) T(\lambda), \qquad T(\lambda) = \begin{pmatrix} a & \bar b \\ b & \bar a \end{pmatrix}, \\ \Delta(\lambda) &= \det T(\lambda) = \frac{W(f_1, \bar f_1)}{W(f_2, \bar f_2)} = \frac{\operatorname{Im} m_1(\lambda_+)}{\operatorname{Im} m_2(\lambda_-)}, \end{aligned} \tag{22.43}$$

where $W(f,g)$ is the Wronskian of the pair (f,g).

Consider vector-rows $\psi_+(\lambda) = (f_1, \bar f_2)$ and $\psi_-(\lambda) = (f_2, \bar f_1)$, which are analytical in the domains M_1 and M_2, respectively. The behaviour of $\psi_\pm(\lambda)$ at infinity will be studied below. We have the relation on E,

$$\begin{aligned} \psi_+(\lambda) &= (f_1(\lambda_+), \overline{f_2(\lambda_-)}), \\ \varphi_-(\lambda) &= (f_2(\lambda_-), \overline{f_1(\lambda_+)}), \end{aligned}$$

since $\tilde\lambda_\pm = \lambda_\mp$. We obtain the boundary condition (cf. (22.17))

$$\psi_+(\lambda) = \psi_-(\lambda) \frac{1}{a(\lambda)} \begin{pmatrix} \Delta(\lambda) & -\overline{b(\lambda)} \\ b(\lambda) & 1 \end{pmatrix}, \qquad \lambda \in E. \tag{22.44}$$

By analogy with (22.18) and (22.19), we introduce the operator

$$f^*(x,\lambda) = \frac{df}{dx} + ikf(x,\lambda) \tag{22.45}$$

and the matrices

$$\begin{aligned} \Psi_+(\lambda) &= \begin{pmatrix} f_1(x,\lambda) & \bar f_2(x,\lambda) \\ f_1^*(x,\lambda) & [\bar f_2(x,\lambda)]^* \end{pmatrix}, \\ \Psi_-(\lambda) &= \begin{pmatrix} f_2(x,\lambda) & \bar f_1(x,\lambda) \\ f_2^*(x,\lambda) & [\bar f_1(x,\lambda)]^* \end{pmatrix}. \end{aligned} \tag{22.46}$$

The matrices (22.46) satisfy the equation

$$\begin{aligned} (\Psi_\pm)_x &= [-ik\sigma_3 + Q(x,t)]\Psi_\pm, \\ Q(x,t) &= \begin{pmatrix} 0 & 1 \\ u - \lambda_{2n} & 0 \end{pmatrix} \end{aligned} \tag{22.47}$$

and the boundary condition

$$\Psi_+(\lambda) = \Psi_-(\lambda)\frac{1}{a(\lambda)}G(\lambda), \qquad \lambda \in E,$$
$$(22.48) \qquad G(\lambda) = \begin{pmatrix} \Delta(\lambda) & -\overline{b(\lambda)} \\ b(\lambda) & 1 \end{pmatrix},$$
$$\Psi_\pm(\lambda) \sim \begin{pmatrix} 1 & 1 \\ 0 & 2ik \end{pmatrix} \exp(-ikx\sigma_3) \qquad \text{as} \quad \lambda \longrightarrow \infty.$$

Note: Evidently, if one of the functions $m_j(\lambda)$ is real on some part of E, the matrix $G(\lambda)$ is triangular in this domain. If both the functions $m_j(\lambda)$ are real, $\frac{1}{a(\lambda)}G(\lambda) = \begin{pmatrix} 0 & 1 \\ 1 & 0 \end{pmatrix}$.

We continue to study coefficients dynamics of problem (22.48).

Consider the two equation systems

$$(22.49') \qquad \psi_x = L(x,t,\lambda)\psi,$$

$$(22.49'') \qquad \psi_t = M(x,t,\lambda)\psi,$$

where L and M are 2×2-matrices. The system (22.49′)–(22.49″) is, in general, incompatible. For the compatibility of this system, it is necessary and sufficient that the Zakharov–Shabat condition

$$(22.50) \qquad L_t - M_x + [L, M] = 0$$

be satisfied. The equation (22.50) is a nonlinear equation with respect to "potentials" L and M. The system (22.49′) and (22.49″) is called the (L, M)-pair for the equation (21.50). In particular, if Eq. (22.50) is a KdV equation, then the (L, M)-pair consists of the Eq. (21.47) and of the equation

$$(\Psi_\pm)_t = [-4ik^3\sigma_3 + R(x,t,k)]\Psi_\pm,$$
$$(22.51) \qquad R(x,t,k) = 4k^2\begin{pmatrix} 0 & 1 \\ \tilde{u} & 0 \end{pmatrix} - 2ik\begin{pmatrix} \tilde{u} & 0 \\ \tilde{u}_x & -\tilde{u} \end{pmatrix} + \begin{pmatrix} -\tilde{u}_x & 2\tilde{u} \\ 2\tilde{u}^2 - \tilde{u}_{xx} & \tilde{u}_x \end{pmatrix},$$
$$\tilde{u}(x,t) = u(x,t) - \lambda_{2n}, \qquad \sigma_3 = \begin{pmatrix} 1 & 0 \\ 0 & -1 \end{pmatrix}$$

(see Chapter 1.7 [40]). Every point of the space of 2×2-matrices belongs to a single integral surface of the system (22.47)–(22.51) (λ is assumed to be fixed and different from the eigenvalues $\eta_1, \dots, \eta_n$ of the operator $L(t)$ and the zone ends $\lambda_0, \dots, \lambda_{2n}$). Consider the integral surface containing the integral curve (22.46) at $t = 0$. The sections $t = \text{const}$ and $x = \text{const}$ of this surface are integral curves of the equations (22.47) and (22.51), respectively. At every moment t, these solutions are denoted by $\tilde{\Psi}_\pm(\lambda)$, and the corresponding matrix elements are denoted by $\tilde{f}_j(x, \lambda)$. These functions coincide with (22.33), for $t = 0$. Therefore, $\tilde{\Psi}_\pm(\lambda)$ is the system (22.51) solution whose initial value is $\Psi_\pm(\lambda)|_{t=0}$. As long as the initial values and right-hand side of Eq. (22.51) are analytic with respect to λ in the domains

$$M_j\setminus\left\{\bigcup_s \eta_s\right\} \qquad (j=1 \quad \text{for} \quad \widetilde{\Psi}_+ \quad \text{and} \quad j=2 \quad \text{for} \quad \widetilde{\Psi}_-),$$

matrices $\widetilde{\Psi}_\pm(\lambda)$ are analytic on the corresponding sheets M_j, with poles at the points η_s and logarithmic singularities at the points λ_k.

As follows from (22.47) and (22.51), the matrices $\widetilde{\Psi}_\pm(\lambda)$ possess the asymptotic behaviour, as $\lambda \longrightarrow \infty$,

$$\widetilde{\Psi}_\pm(\lambda) \sim \begin{pmatrix} 1 & 1 \\ 0 & 2ik \end{pmatrix} \exp(-ikx - 8ik^3t)\sigma_3. \tag{22.52}$$

The coefficients dynamics of (22.48)[1] may be obtained from (22.51). Fix the arbitrary values x_1 and x_2 and consider the integral curves $\widetilde{\Psi}_\pm(\lambda)_{|x=x_j}, j=1,2$. Using (22.9), we calculate the values $a(\lambda)$ and $b(\lambda)$ for every t. The operator

$$\begin{aligned} &\widetilde{T}(\lambda) = K_\lambda[\omega(t)], \\ &\omega(t) = (\tilde{u}(x_1,t), \tilde{u}'_x(x_1,t), \tilde{u}''_{xx}(x_1,t); \\ &\qquad \tilde{u}(x_2,t), \tilde{u}'_x(x_2,t), \tilde{u}_{xx}(x_2,t)) \end{aligned} \tag{22.53}$$

has a triangle form. Let

$$\Psi^0_\pm(\lambda) = \widetilde{\Psi}_\pm(\lambda)\exp(ikx + 8ik^3t)\sigma_3. \tag{22.54}$$

We obtain the boundary problem

$$\begin{aligned} &\Psi^0_+(\lambda) = \Psi^0_-(\lambda)e^{(-ikx-8ik^3t)\sigma_3}\frac{\widetilde{G}(\lambda)}{\overline{a(\lambda}}e^{(ikx+8ik^3t)\sigma_3}, \\ &\lambda \in E, \qquad \Psi^0_\pm(\lambda)\begin{pmatrix} 1 & -(2ik)^{-1} \\ 0 & (2ik)^{-1} \end{pmatrix} \longrightarrow 1 \qquad \text{as} \quad \lambda \longrightarrow \infty. \end{aligned} \tag{22.55}$$

in the class of matrices which are multiples of the divisor $\gamma = \sum \eta_k$ and have logarithmic singularities at the points $\lambda_0, \dots, \lambda_{2n}$.

In some cases, problem (22.55) may be solved effectively (see Appendix 3).

Solving problem (22.55), we have the formula for the potential and its derivatives

$$\begin{pmatrix} -\tilde{u}'_x(x,t) & 2\tilde{u}(x,t) \\ 2\tilde{u}^2(x,t) - \tilde{u}''_{xx}(x,t) & u'_x(x,t) \end{pmatrix} = \widetilde{\Psi}_t(\lambda_{2n})\widetilde{\Psi}^{-1}(\lambda_{2n}). \tag{22.56}$$

---

[1]Here it is necessary to use the coefficients of

$$\tilde{f}_1(x,\lambda_+) = \tilde{a}(\lambda)\tilde{f}_2(x,\lambda_-) + \tilde{b}(\lambda)\overline{\tilde{f}_2(x,\lambda_-)} \tag{22.41'}$$

instead of a, b. The corresponding matrices are denoted by $\widetilde{G}, \widetilde{T}$.

Assuming $x = x_1, x_2$ in this equation, we obtain the nonlinear equation

$$\omega(t) = H[\omega(t)], \tag{22.57}$$

where H is a triangular operator. Solving this equation, we may calculate the values $a(\lambda), b(\lambda)$ by the formula (22.53)

Consider in more detail the case of a periodic potential [9], [40]. Taking into account the branches of the function $\sqrt{R(\lambda)}$ and Eqs. (22.37), (22.42), we conclude that in this case, $b(\lambda) \equiv 0$, and $a(\lambda) \equiv 1$. Therefore, we have the "reflectionless" situation,

$$\tilde{\Psi}_+(\lambda) = \tilde{\Psi}_-(\lambda), \qquad \lambda \in E. \tag{22.58}$$

The function

$$B(\lambda) = \begin{cases} \tilde{f}_{1}, (x, \lambda), & \lambda \in M_1 \\ \tilde{f}_{2}, (x, \lambda), & \lambda \in M_2 \end{cases} \tag{22.59}$$

is called the Baker–Akhiezer function. We have

$$B(\lambda) \sim \exp\{-ikx - 8ik^3 t\} \qquad \text{as} \quad \lambda \longrightarrow \infty. \tag{22.60}$$

The Baker–Akhiezer function poles are determined by the initial values. As is seen from Eq. (22.37), only one of the functions, $m_j(\lambda)(j = 1, 2)$, has a pole at the point $\xi_k(k = 1, \dots, n)$ according to chosen branches of the values $\sqrt{-R(\xi_k)}$. This means that the function $B(\lambda)$ has only one pole, $\eta_k(\xi_k, j)(j = 1, 2, k = 1, \dots, n)$, over the point ξ_k (j means the number of sheets of the surface M) and hence has n poles on M. Therefore, the function $B(\lambda)$ can be represented by the form

$$B(\lambda) = \beta_j(\lambda)\varphi_j(\lambda), \qquad \lambda \in M_j, \qquad j = 1, 2, \tag{22.61}$$

where the functions $\varphi_j(\lambda)$ are analytic in $M_j, \varphi_j(\infty) = 1$, and

$$\beta_j(\lambda) = \exp\{-ikx - 8ik^3 t\} \prod_{\eta_l \in M_j} \frac{k + (-1)^j \xi}{k - \sqrt{\xi_l - \lambda_{2n}}}, \tag{22.62}$$

where ξ is an arbitrary point belonging to the lower halfplane, $\operatorname{Im} \xi < 0$. We obtain the scalar Riemann problem of the index n on the surface M,

$$\varphi_1(\lambda) = \frac{\beta_2(\lambda)}{\beta_1(\lambda)}\varphi_2(\lambda), \qquad \lambda \in E, \qquad \varphi_j(\infty) = 1 \tag{22.63}$$

(see §8). Calculate the values (6.19). As long as $\lambda = \infty$ is a boundary point of the domains M_j, we take into account the semiresidues

$$\int_{E^*} \ln \beta_j(\lambda) dw_s(\lambda) = (-1)^j \frac{i}{2}[xw_s'(\infty) + 4tw_s'''(\infty)]. \tag{22.64}$$

Here

$$w_s'(\infty) = \frac{dw_s(\lambda)}{d\varsigma}|_\varsigma = 0',$$
$$w_s'''(\infty) = \frac{d^3 w_s(\lambda)}{d\varsigma^3}|_\varsigma = 0', \qquad \varsigma = \frac{1}{k},$$

and E^* is the boundary contour of the sheets M_j,

$$E^* = \sum_{j=0}^{n} E_j^*, \qquad E_j^* = (\lambda_{2j}, \lambda_{2j+1})^*, \qquad j = 0, \ldots, n-1,$$
$$E_n^* = (\lambda_{2n}, \infty)^*,$$

where any component E_j^* is the corresponding segment passed twice. The passing direction is chosen such that the domain M_1 is situated to the left of E^*. Draw the cuts from the points η_l to the boundary points $q_l (l = 1, \ldots, n)$. We obtain the relation

(22.65)
$$-\int_{E^*} \sum_{\eta_l \in M_1} \ln \frac{k-\xi}{k - \sqrt{\xi_l - \lambda_{2n}}} dw_s + \int_{E^*} \sum_{\eta_l \in M_2} \ln \frac{k+\xi}{k - \sqrt{\xi_l - \lambda_{2n}}} dw_s$$
$$= -2\pi i \sum_{r=1}^{n} \int_{q_r}^{\eta_r} dw_s, \qquad s = 1, \ldots, n.$$

Whence it follows that

$$l_s = ixw_s'(\infty) + 4itw_s'''(\infty) + \sum_{r=1}^{n} w_s(\eta_r), \tag{22.66}$$

where $w_s(\lambda)$ are the normalized Abelian integrals of the first kind. From Eq. (8.15) it follows that

(22.67)
$$\varphi_j(\lambda) = \theta(w_s(\lambda) - k_s - l_s) \exp\left\{ \frac{x}{2\pi} \sum_{l=0}^{n} \int_{E_l^*} [k_1(\tau) - k_2(\tau)]\, M^*(\tau_1 \lambda) d\tilde{\imath} + \Phi_j(\lambda) \right\},$$
$$\lambda \in M_j.$$

Here

$$k_j(\lambda) = \sqrt{\lambda - \lambda_{2n}}, \qquad \lambda \in M_j, \qquad j = 1, 2,$$

and the functions $\Phi_j(\lambda)$ determined by Eq. (8.15) are independent of x.

Note that in this case, $\alpha = 0$. Indeed, zeros of a solution of (22.63) coincide with zeros of $B(\lambda)$. As is known, this function has n zeros, $\varepsilon_1, \ldots, \varepsilon_n$ (one zero in every lacuna) [25, 40]. Hence $\dim L(\sum_1^n \varepsilon_j) = 1$ (see Appendix 1), and

$$\theta(w_s(\lambda) - l_s - k_s) \not\equiv 0. \tag{22.68}$$

Let

$$\chi = \frac{d\ln B(\lambda)}{dx}. \tag{22.69}$$

The function χ satisfies the Riccati equation

$$\frac{d\chi}{dx} + \chi^2 - u + \lambda = 0. \tag{22.70}$$

We have the decompositon at infinity

$$\chi(\lambda) = -ik + \sum_{n=1}^{\infty} \frac{\chi_n(x)}{k^n}. \tag{22.71}$$

Then from Eq. (22.70) it follows that

$$u(x,t) = -2i\chi_1(x) + \lambda_{2n} - i. \tag{22.72}$$

Substitute the variable $\varsigma = k^{-1}$for k in Eq. (22.71). Then we get the relation

$$\chi_1(x) = \frac{d[\chi(\lambda) + ik]}{d\varsigma}_{|\varsigma=0}. \tag{22.73}$$

We have

$$\chi(\lambda) = -ik + \frac{d\ln\theta(w_s(\lambda) - k_s - l_s)}{dx} + \frac{1}{2\pi}\sum_{l=0}^{n}\int_{E_l^*}[k_1(\tau) - k_2(\tau)]M^*(\tau,\lambda)d\tau. \tag{22.74}$$

Note that

$$\frac{\partial(w_s(\lambda) - k_s - l_s)}{\partial x} = i\frac{\partial(w_s(\lambda) - k_s - l_s)}{\partial\varsigma}_{|\varsigma=0} = iw_s'(\infty). \tag{22.75}$$

Hence,

$$u(x,t) = -2i\chi_1(x) + \lambda_{2n} - i = -2i\frac{d[\chi(\lambda) + ik]}{d\varsigma}_{|\varsigma=0}$$
$$+\text{const} = -2i\frac{d^2\ln\theta(w_s(\lambda) - k_s - l_s)}{dx\partial\varsigma}_{|\varsigma=0}$$
$$+\text{const} = -2i\frac{d^2\ln\theta(w_s(\lambda) - k_s - l_s)}{dx^2}_{|\lambda=\infty} + \text{const}.$$

We normalize the Abelian integrals of the first kind by the condition

$$w_s(\infty) = 0, \qquad s = 1, \ldots, n.$$

Then we obtain the famous Its–Matveev formula

$$(22.76)\qquad u(x,t) = -2\frac{d^2}{dx^2}\ln\theta(-ixw_s'(\infty) - 4itw_s'''(\infty) - \sum_{r=1}^{n} w_s(\eta_r) - k_s) + \text{const}$$

for a finite-zone periodical potential.

Let now M be the same surface and ξ_s be fixed points belonging to the lacunas $(\lambda_{2s-1}, \lambda_{2s})$, $s = 1, \dots n$. Consider the Baker–Akhiezer function $B(\lambda)$ defined by the asymptote at infinity (22.60), with poles of the first order at the n points $\eta_s(\xi_s, j)$, $j = 1$ or 2, and satisfying the condition $B(\lambda)_{|x=0,t=0} = 1$. In this case, the function (22.76) is a solution of the KdV equation. This potential is finite-zone and reflectionless. As was shown by Novikov, who discovered this potential class, $u(x,t)$ is an almost-periodic function with respect to x and t [9, 40].

Indeed, the vector $A(p)(A_1(p), \dots, A_n(p))$,

$$(22.77)\qquad A_s(p) = w_s(p) - k_s - l_s,$$

defines a point of the Jacobi variety J of the surface M. The values $A_s(p)$ linearly depend on x and t. Hence the point A moves on J periodically or almost periodically as a function of x and t.

§23 The Landau–Lifschitz Equation

A. *Fast-Decreasing Potentials*

In this section, we follow the Zakharov–Shabat method [a,b] called the "Riemann problem method."

We consider the equation

$$(23.1)\qquad \begin{aligned} &\mathbf{S}_t = \mathbf{S}\times\mathbf{S}_{xx} + \mathbf{S}\times J\mathbf{S}, && |\mathbf{S}| = 1,\\ &J = \operatorname{diag}(J_1, J_2, J_3), && J_1 \leqslant J_2 \leqslant J_3. \end{aligned}$$

This equation describes the spin waves in ferromagnets (Lakshmanan [a], Takhtajan [a], Zakharov & Takhtajan [a], Sklyanin [a], Borovik [a], Mikhailov [d], Rodin [r,s], Borisov [a], and the book [23]). The equation (23.1) is a compatibility condition for the systems

$$(23.2)\qquad i\varphi_x = L\varphi, \qquad L(x,t) = \sum_{\alpha=1}^{3} w_\alpha(\lambda) S_\alpha(x,t)\sigma_\alpha,$$

$$(23.3)\qquad i\varphi_t = M\varphi, \qquad M(x,t) = \sum_{\alpha=1}^{3} w_\alpha(\lambda) S_\beta S_{\gamma x}\sigma_\alpha, \varepsilon^{\alpha\beta\gamma} - \sum_{\alpha=1}^{3} w_\beta(\lambda) w_\gamma(l) S_\alpha\sigma_\alpha |\varepsilon^{\alpha\beta\gamma},$$

where σ_α are the Pauli matrices

$$\sigma_1 = \begin{pmatrix} 0 & 1 \\ 1 & 0 \end{pmatrix}, \qquad \sigma_2 = \begin{pmatrix} 0 & -i \\ i & 0 \end{pmatrix}, \qquad \sigma_3 = \begin{pmatrix} 1 & 0 \\ 0 & -1 \end{pmatrix}.$$

$\varepsilon^{\alpha\beta\gamma}$ is a completely antisymmetric tensor of the third rank, $\varepsilon^{123} = 1$, and $w_\alpha(\lambda)$ are elliptic functions in the rectangle $R\{|\operatorname{Re}\lambda| \leqslant 2K, |\operatorname{Im}\lambda| \leqslant 2K'\}$,

$$\begin{aligned} w_1(\lambda) &= \frac{\rho}{\operatorname{sn}(\lambda,k)}, \qquad w_2(\lambda) = \frac{\rho\operatorname{dn}(\lambda,k)}{sn(\lambda,k)}, \\ w_3(\lambda) &= \frac{\rho\operatorname{cn}(\lambda,k)}{sn(\lambda,k)}, \qquad k = \sqrt{\frac{J_2-J_1}{J_3-J_1}}, \qquad \rho = \frac{1}{2}\sqrt{J_3-J_1}. \end{aligned} \tag{23.4}$$

If $J_1 = J_2 = J_3$,

$$w_\alpha(\lambda) = \frac{1}{\lambda}, \qquad \alpha = 1, 2, 3,$$

and R degenerates into the complex λ-plane.

We consider the Cauchy problem for the equation (23.1) and suppose that $\mathbf{S}(x,0)$ is sufficiently smooth, and $\mathbf{S}(x,0) \to (0,0,1)$ sufficiently fast as $|x| \to \infty$. The Jost solutions of the equation (23.2) are defined by the asymptotes

$$f_\pm(x,\lambda) = \exp[-iw_3(\lambda)x\sigma_3] + o(1) \qquad \text{as} \quad x \to \pm\infty \tag{23.5}$$

and are connected by the transition matrix $T(\lambda)$,

$$f_+(x,\lambda) = f_-(x,\lambda)T(\lambda), \qquad \operatorname{Im}\lambda = 0, 2K'. \tag{23.6}$$

It has the form (Sklyanin [a])

$$T(\lambda) = \begin{pmatrix} a(\lambda) & -\overline{b(\bar\lambda)} \\ b(\lambda) & \overline{a(\bar\lambda)} \end{pmatrix}, \tag{23.7}$$

$$\begin{aligned} a(\lambda+2K) &= a(\lambda), \qquad & b(\lambda+2K) &= b(\lambda), \\ a(\bar\lambda+2iK') &= \overline{a(\lambda)}, \qquad & b(\bar\lambda+2iK') &= -\overline{b(\lambda)}. \end{aligned} \tag{23.8}$$

From $SpL = 0$, it follows that $\det f_\pm(x,\lambda)$ are independent of x, $\det f_\pm(x,\lambda) = 1$, and

$$\det T(\lambda) = |a(\lambda)|^2 + |b(\lambda)|^2 = 1, \qquad \operatorname{Im}\lambda = 0, 2K'. \tag{23.9}$$

The matrices

$$\varphi_\pm(x,\lambda) = f_\pm(x,\lambda)\exp iw_3(\lambda)x\sigma_3 \tag{23.10}$$

satisfy the equation

$$i(\varphi_\pm)_x = L\varphi_\pm - w_3(\lambda)\varphi_\pm\sigma_3. \tag{23.11}$$

We get the integral equation

$$\varphi_\pm(x,\lambda) - i\int_x^{\pm\infty} \exp\{-iw_3(\lambda)(x-y)\sigma_3\}\cdot[L(y,t) - w_3(\lambda)\sigma_3]\varphi_\pm(y,\lambda)\exp\{iw_3(\lambda)(x-y)\sigma_3\}dy = 1. \tag{23.12}$$

This representation provides the analytic continuation of the first column φ_+^1 of the matrix φ_+ and the second column φ_-^2 of the matrix φ_- into the upper half R_+ of the rectangle R. The second column φ_+^2 and the first column φ_-^1 of the matrices φ_+ and φ_- are analytically continued into the lower half R_- of R (see, for example, Rodin [s]).

Assume

$$\psi_+(\lambda) = (\varphi_+^1, \varphi_-^2), \qquad \psi_-(\lambda) = (\psi_-^1, \varphi_+^2). \tag{23.13}$$

These matrices are analytic in the domains $R_\pm$, respectively. As long as $\det\psi_+(\lambda) = a(\lambda)$, this function is also analytic in R_+. Zeros $\lambda_1, \ldots, \lambda_n$ of $a(\lambda)$ form the discrete spectrum, since $f_+(\lambda_j) = \beta_j f_{(\lambda_j)}$ is a solution decreasing for $|x| \to \infty$.

We shall introduce the contour

$$\Gamma = \Gamma_1 \cup \Gamma_2, \qquad \Gamma_1 = R \cap \{\mathrm{Im} = 0\}, \qquad \Gamma_2 = R \cap \{\mathrm{Im}\,\lambda = 2K'\}.$$

We get the relation on Γ,

$$\begin{gathered} \psi_+(\lambda) = (\phi_+^1, \phi_-^2) = \frac{1}{\bar a(\lambda)}(\phi_-^1 + b\phi_+^2 \exp 2iw_3(\lambda)x, \\ \bar b\phi_-^1 \exp(-2iw_3(\lambda)x) + \varphi_-^2 = \psi_-(\lambda)\frac{G(x,t,\lambda)}{\bar a(\lambda)}, \\ G(x,t,\lambda) = e^{iw_3(\lambda)x\sigma_3}\begin{pmatrix} 1 & \overline{b(\lambda)} \\ b(\lambda) & 1\end{pmatrix} e^{-iw_3(\lambda)x\sigma_3}. \end{gathered} \tag{23.14}$$

As long as $\psi(\lambda+4Kn) = \psi(\lambda+4iK'm) = \psi(\lambda)$, we obtain the Riemann boundary problem on the torus corresponding to the rectangle R.

For simplicity, we limit ourselves to the case $a(\lambda) \neq 0$ in R_+. Therefore, we suppose that the discrete spectrum is absent; for the general case and for the construction of the soliton solutions ($b = 0$), see Rodin [s].

A solution of the boundary problem (23.14) is determined up to a constant matrix factor depending on x and t. In order to separate the partial solution, we use the problem symmetry.

Using (23.11) and the behaviour of the solution at infinity, one can verify that the doubly-periodic solution $\varphi_\pm(\lambda)$ has the following symmetries,

$$\psi_\pm(\lambda + 2K) = \sigma_3\psi_\pm(\lambda)\sigma_3, \tag{23.15$_1$}$$

$$\psi_\pm(\bar\lambda + 2iK') = \sigma_3\overline{\psi_\pm(\lambda)}\sigma_3, \tag{23.15$_2$}$$

$$\psi_\pm(\bar\lambda) = (\psi_\mp(\lambda))^+, \tag{23.15$_3$}$$

where ψ^+ is the Hermitian conjunction symbol. These properties fix the solution. Indeed, if $\psi \pm (\lambda)$ and $C\psi_\pm(\lambda)$ are two solutions satisfying (23.15), then

$$\sigma_3 C \sigma_3 = C, \qquad \sigma_3 \bar{C} \sigma_3 = C, \qquad C^+ = C^{-1}. \tag{23.16}$$

This means that $C = \pm 1$ or $\pm\sigma_3$.

If $\psi_\pm(\lambda)$ is a solution of the problem (23.14), the solution possessing the properties (23.15) may be obtained by averaging of the symmetry group. Let $\{g_\alpha\}$ be some set of motions of R generating the finite transformation group $\{\mathfrak{G}\}$, and let $T(g)$ be a 2×2-matrix representation of $\{\mathfrak{G}\}$ (reduction group, Mikhailov [c]). We assume

$$\psi(\lambda) = \frac{1}{n} \sum_g T^{-1}(g)\psi(g)T(g), \tag{23.17}$$

where n is the order of the group $\{\mathfrak{G}\}$.

Here $g_1\lambda = \lambda + 2K$, $g_2\lambda = \bar{\lambda} + 2iK'$. We do not take into account (23.15$_3$). By (23.16) the solution is defined up to a diagonal matrix $C(x,t)$. Because of (23.22) the problem solution $S(x,t)$ is independent of $C(x,t)$. As long as $|b| < 1$ and the operator A is an orthogonal projector, a partial solution of the boundary problem (23.14) may be obtained by the method of Appendix 3.

The kernel of the Cauchy-type integral on the torus has the form

$$M(\tau, \lambda) = \varsigma(\tau - \lambda) - \varsigma(\tau - iK') + \varsigma(\lambda - K - iK') + \varsigma(K). \tag{23.18}$$

The dynamics of the coefficients of (23.14) follows from (23.3) for $x \longrightarrow -\infty$. From (23.7), we have

$$f_+(x,\lambda) \sim \exp(-i\omega_3(\lambda)x\sigma_3)T(\lambda) \qquad \text{as} \quad x \longrightarrow -\infty. \tag{23.19}$$

As seen from the Zakharov–Shabat equation (22.50), the operator M is defined up to the unit operator with scalar coefficient. We get the equation, as $x \longrightarrow -\infty$,

$$\dot{f}_+(x,\lambda) = [2i\omega_1(\lambda)\omega_2(\lambda)\sigma_3 + \mu E\} f_+, \qquad E = \begin{pmatrix} 1 & 0 \\ 0 & 1 \end{pmatrix} \tag{23.20}$$

We choose the coefficient $\mu = -2i\omega_1(\lambda)\omega_2(\lambda)$. Then, by substituting (23.18) in (23.20), we get the GGKM equations

$$\dot{a}(\lambda) = 0, \qquad \dot{b}(\lambda) = -4i\omega_1(\lambda)\omega_2(\lambda)b(\lambda). \tag{23.21}$$

The solution of the inverse scattering problem may be obtained from Eq. (23.11). As long as the functions $\psi \pm (\lambda)$ are regular at $\lambda = 0$, the residue of both sides of Eq. (23.11) are equal to zero, and we get the formula

$$S(x,t) = \sum_{\alpha=1}^{3} \sigma_\alpha S_\alpha = \begin{pmatrix} S_3 & S_1 - iS_2 \\ S_1 + iS_2 & -S_3 \end{pmatrix} = \psi_\pm(0)\sigma_3\psi_\pm^{-1}(0). \tag{23.22}$$

Details of the construction, the case of soliton solutions, and so on are treated in the papers by Mikhailov [d] and Rodin [s].

B. *Reflection Finite-Zone Potentials*

The general method of the calculation of the exact solutions (23.11) in the framework of the finite-zones integration was proposed by Bogdan & Kovalev [a], Bobenko [a,b], Bikbaev & Bobenko [a], Bikbaev, Bobenko & Its [a]. Here we follow, in general, the later paper.

Let $\Psi(\lambda)$ be a meromorphic 2×2-matrix in R which is double-periodic and has the asymptote at $\lambda = 0$,

$$\Psi(\lambda) = \left(\sum_{j=0}^{\infty} \Phi_j(x,t)\lambda^j\right) \exp(-\frac{i\rho x}{\lambda} + \frac{2i\rho^2 t}{\lambda^2})\sigma_3. \tag{23.23}$$

The matrix $\Psi(\lambda)$ has to satisfy the following restrictions. The matrix $\Psi(\lambda)$ possesses the singularities at the points $\lambda_1, \ldots, \lambda_{2g} \in R$, independent of x and t,

$$\Psi(\lambda) = \widehat{\Psi}_j(\lambda)(\lambda - \lambda_j)^{\begin{pmatrix} 0 & 0 \\ 0 & 1/2 \end{pmatrix}} \begin{pmatrix} -1 & -1 \\ 1 & 1 \end{pmatrix}. \tag{23.24}$$

It is necessary that on the cuts $\Gamma(\lambda_{2j-1}, \lambda_{2j})(j = 1, \ldots, g)$, the conditions

$$\Psi_+(\lambda) = \Psi_-(\lambda)\sigma_1 \tag{23.25}$$

be satisfied. It is clear that the matrix $\Psi(\lambda)$ is single-valued on the two-sheeted covering surface $\widehat{R}$ over the torus R with the branch points $\lambda_1, \ldots, \lambda_{2g}$.

At the points $\mu_1, \ldots, \mu_n \in R$, the matrix $\Psi(\lambda)$ has to be represented by

$$\Psi(\lambda) = \widehat{\Psi}_k(\lambda)(\lambda - \mu_k)^{\begin{pmatrix} -1 & 0 \\ 0 & 0 \end{pmatrix}}, \qquad k = 1, \ldots, n, \tag{23.26}$$

where $\widetilde{\Psi}_k(\lambda)$ are nondegenerate.

Further, the following reduction conditions must be satisfied

$$\begin{aligned} \sigma_3\Psi(\bar{\lambda} + 2iK')\sigma_3 &= \Psi(\lambda), \\ \sigma_3\Psi(\lambda + 2K)\sigma_3 &= \Psi(\lambda). \end{aligned} \tag{23.27}$$

Finally, on the contours

$$\Gamma_{ij}\{|\operatorname{Re}\lambda| \leqslant 2K, \operatorname{Im}\lambda = 0\}$$

and

$$\Gamma_{2j}\{|\operatorname{Re}\lambda| \leqslant 2K, \operatorname{Im}\lambda = 2K'\}$$

belonging to the corresponding sheets $R_j (j = 1, 2)$ of the surface $\widehat{R}$, the following boundary condition has to be satisfied

$$\Psi_+(\lambda) = \Psi_-(\lambda)\frac{G(\lambda)}{\overline{a}(\lambda)},$$

$$G(\lambda) = \exp[-iw_3(\lambda)x\sigma_3]\begin{pmatrix} 1 & \overline{b(\lambda)} \\ b(\lambda) & 1 \end{pmatrix}\exp[iw_3(\lambda)x\sigma_3], \tag{23.28}$$

$$|a(\lambda)|^2 + |b(\lambda)|^2 = 1,$$

$$\lambda \in \Gamma = \bigcup_{k,j=1}^{2} \Gamma_{kj}.$$

The matrix $\Psi(\lambda)$ is a solution of the equations (23.2), (23.3). Indeed, the matrixes $\Psi_x\Psi^{-1}$ and $\Psi_t\Psi^{-1}$ are single-valued on R, continuous on the contours $\Gamma_{kj}(k, j = 1, 2)$, and holomorphic at the points $\lambda_1, \ldots, \lambda_{2g}, \mu_1, \ldots, \mu_n$. The following construction may be carried out by the standard scheme (see the paper by Bikbaev, Bobenko & Its [a], where it is supposed that $b \equiv 0$).

In conclusion, we note that instead of the boundary condition (23.28), one can assume that the matrix $\Psi(\lambda)$ is not analytic but satisfies the differential equation

$$\bar{\partial}\Psi = \Psi(\lambda)A(\lambda)$$

on $\widehat{R}$, since the matrices $\Psi_x\Psi^{-1}$ and $\Psi_t\Psi^{-1}$ are analytic (see §24).

§24 Riemann–Hilbert and Related Problems

A. *D-BAR Problem*

In the recent papers by Beals & Coifman [a,b], Fokas & Ablowitz [a], Ablowitz, Bar Yaacov & Fokas [a], and others, the fundamental applications of CBV systems (see Chapter 5; it is also called the *D-BAR* problem) to the inverse scattering problem, in particular for a multi-dimensional case, were investigated. We state one approach to this problem, following the paper by Beals & Coifman [b] with some modification. We limit ourselves to aspects illustrating the methods described in this book.

Consider the system

$$\partial_x\psi = [\lambda J + Q(x)]\psi, \qquad \partial_x \equiv \frac{\partial}{\partial x}, \tag{24.1}$$

where $J = \operatorname{diag}(J_1, J_2)$ is a matrix constant and $Q(x)$ is a 2×2-off-diagonal matrix, continuous and bounded at infinity.

The solution of (24.1) is found in the form

$$\psi(x, \lambda) = \varphi(x, \lambda)\exp\lambda Jx. \tag{24.2}$$

We get the equation

$$D_\lambda\varphi = Q\varphi, \qquad D_\lambda = \partial_x - \lambda\operatorname{ad}J, \qquad \operatorname{ad}J = [J, \cdot] \tag{24.3}$$

If φ_1 and φ_2 are solutions of Eq. (24.1) and φ_1 is invertible, then

$$D_\lambda(\varphi_1^{-1}\varphi_2) = 0. \tag{24.4}$$

Let $\varphi(x,\lambda)$ have poles at the points $\lambda = \mu_k (k = 1,\ldots,n)$ and at the branch points $\lambda = \lambda_j (j = 1,\ldots,m)$ at which $\varphi(x,\lambda)$ is represented in the form

$$\varphi(x,\lambda) = \hat{\varphi}_j(x,\lambda)(\lambda-\lambda_j)^{T_j}C_j, \qquad j = 1,\ldots,m. \tag{24.5}$$

Here the matrices $\hat{\varphi}_j(x,\lambda)$ are degenerate, T_j, C_j are independent of x and λ, and T_j are diagonal and rational.

Therefore, $\varphi(x,\lambda)$ is single-valued on some multisheeted surface M over the λ-plane. It can be supposed that λ is a point of a Riemann surface M_0 (in §23, M_0 is a torus). In this case M is a covering surface over M_0 (see Zakharov & Mikhailov [a]).

As long as the operators D_λ and $\bar{\partial} = \frac{\partial}{\partial\bar{\lambda}}$ commute, the matrix $\bar{\partial}\varphi(x,\lambda)$ is also a solution of Eq. (24.3), and hence

$$\bar{\partial}\varphi = \varphi a, \qquad D_\lambda a = 0. \tag{24.6}$$

From the equation $D_\lambda a = 0$, it follows that

$$a(x,\lambda) = \exp(\lambda Jx)w(\lambda)\exp(-\lambda Jx). \tag{24.7}$$

As long as the matrices $\exp(\lambda Jx)$ and $(\lambda-\lambda_j)^{T_j}$ commute, $w(\lambda)$ is an arbitrary 2×2 matrix on the surface of M. Evidently,

$$\psi(x,\lambda) = \exp(\lambda Jx) + o(1) \quad \text{as} \quad \lambda\to\infty, \tag{24.8}$$

and hence

$$\varphi(x,\lambda) = 1 + o(1) \quad \text{as} \quad \lambda\to\infty. \tag{24.9}$$

From (24.6) we get the integral equation

$$\varphi(x,\lambda) + \frac{1}{\pi}\iint_M \varphi(x,\tau)a(x,\tau)M(\tau,\lambda)d\sigma_\tau = \Phi(\lambda), \tag{24.10}$$

where $\Phi(\lambda)$ is an analytic matrix on M which is a multiple of the divisor $-\Delta$, $\Delta = \delta + \sum_k \mu_k$, where $\delta = \sum_j p_j$ is a characteristic divisor of the Cauchy kernel $M(\tau,\lambda)$, and $\Phi(\infty) = 1$. If the points $\mu_1,\ldots,\mu_n$ are absent, $\Phi(\lambda)\equiv 1$, since $\dim\delta = 1$.

The matrix $w(\lambda)$ plays the role of the scattering data. The reflectionless case corresponds to $w(\lambda)\equiv 1$.

It can be verified that

$$\begin{aligned} Q(x) &= [D_\lambda, T](\varphi a), \\ Tf &= \frac{1}{\pi}\iint_M f(\tau)M(\tau,\lambda)d\sigma_\tau. \end{aligned} \tag{24.11}$$

Let there be two independent variables, $x_1 = x$ and $x_2 = t$, and

$$D^j_\lambda = \partial_{x_j} - \lambda \operatorname{ad} J_j, \qquad j = 1, 2, \tag{24.12}$$

J_j are diagonal constant matrices. The compatability condition for the systems

$$D^j_\lambda \varphi = Q^j \varphi, \qquad j = 1, 2 \tag{24.13}$$

is the Zakharov–Shabat equation

$$D^2_\lambda Q^1 - D^1_\lambda Q^2 + [Q^1, Q^2] = 0. \tag{24.14}$$

A solution of the system $D^j_\lambda a = 0, j = 1, 2$ has the form

$$a(x, t, \lambda) = \exp \lambda (J_1, x + J_2 t) w(\lambda) \exp \lambda(-J_1 x - J_2 t), \tag{24.15}$$

and the solution of the inverse problem has the form

$$Q^j = [D^j_\lambda, T](\varphi a). \tag{24.16}$$

One can show that it has the form

$$Q^j = [J_j, q], \qquad q(x, t) = \iint_M \varphi(x, t, \tau) a(x, t, \tau) d\sigma_\tau. \tag{24.17}$$

B. *The Dressing Method*

In the some cases the Zakharov–Shabat "dressing method" [a,b] leads to the Riemann problem on Riemann surfaces (see Mikhailov [c], Bobenko [a]). The Riemann surfaces generated by the general scheme were proposed by Zakharov & Mikhailov [a]. We follow this paper for the case of hyperelliptic surfaces.

Let $\widehat{M}$ be the hyperelliptic surface of the genus g determined by the equation

$$w^2 = R_{2g}(\lambda), \qquad R_{2g}(\lambda) = \lambda^{2g} + a_{2g-1}\lambda^{2g-1} + \cdots + a_0. \tag{24.18}$$

Introduce the projective coordinate $w = Z_0/Z_1$, $\lambda^k = Z_{k+1}/Z_k$, $1 \leqslant k \leqslant g$. The algebraic curve $\widehat{M}$ is represented as the quadric intersection

$$\begin{gathered} Z_0^2 = Z_{g+1}^2 + a_{2g-1} Z_{g+1} Z_g + \cdots + a_0 Z_1^2, \\ Z_j Z_k - Z_l Z_s = 0, \qquad j + k = l + s, \qquad 1 \leqslant j, k, l, s \leqslant g + 1. \end{gathered} \tag{24.19}$$

The Zakharov–Shabat equation

$$L_t - M_x + [L, M] = 0 \tag{24.20}$$

for the system

$$\begin{aligned} \Psi_x &= L(x, t, \lambda)\Psi, \\ \Psi_t &= M(x, t, \lambda)\Psi, \end{aligned} \tag{24.21}$$

where $\lambda \in \widehat{M}$ and L, M are 2×2 matrices, and

$$L = L_0 + \sum_{k=1}^{n+1} \frac{Z_k}{Z_0} L_k, \qquad M = M_0 + \sum_{k=1}^{n+1} \frac{Z_k}{Z_0 k} M_k, \tag{24.22}$$

takes the form

$$\begin{aligned} &\frac{\partial L_k}{\partial t} - \frac{\partial M_k}{\partial x} = [L_k, M_0] + [L_0, M_k], \\ &\frac{\partial L_0}{\partial t} - \frac{\partial M_0}{\partial x} = [L_0, M_0] + [L_{n+1}, M_{n+1}], \\ &\sum_{s+l=j} [L_s, M_l] = a_{j-2}[L_{n+1}, M_{n+1}], \qquad s, l \geqslant 1, \qquad j \leqslant 2g-1. \end{aligned} \tag{24.23}$$

The system (24.23) is compatible. Its partial solution is

$$L_i = \alpha_i \sigma_3, \qquad M_i = \beta_i, \sigma_3. \tag{24.24}$$

The dressing procedure for this solution ([40], chapter 3) leads to a 2×2 matrix Riemann problem on M.

C. *The Riemann–Hilbert Problem*

Consider on the complex λ-plane the system of equations

$$\frac{d\psi}{d\lambda} = \psi(\lambda) \sum_{j=1}^{m} \frac{A_j}{\lambda - \lambda_j}. \tag{24.25}$$

Here $A_j (j = 1, \ldots, m)$ are constant $n \times n$-matrices, and λ_j are given points. A fundamental matrix of the system (24.25) is multivalued. Indeed, let the point λ move on a path $|\lambda - \lambda_j| = \varepsilon$ encircling the point λ_j. Then the solution takes an increment, and its new value $T_j(\psi)$ is equal to

$$T_j(\psi) = G_j \psi(\lambda), \qquad j = 1, \ldots, m, \tag{24.26}$$

where G_j is a constant matrix. The transformations $\{T_j\}$ generate the monodromy group of Eq. (24.25). The points $\lambda_1, \ldots, \lambda_m$, at which the coefficients of the equation have poles, are called *singular points* of the equation. If the pole order is equal to unity (as in (24.25)), the corresponding singular point is called a *Fuchsian singularity*. The substitution $\tau = \lambda^{-1}$ shows that infinity is a Fuchsian singular point if $\psi^{-1} \frac{d\psi}{d\lambda}$ has zero of the first order at infinity. Therefore, the system whose singular points are Fuchsian has the form (24.25).

The problem formulated by Riemann and known as the 21-Hilbert problem (now it is called the Riemann–Hilbert problem) is to determine the coefficients of the system (24.25) for the given $\lambda_1, \ldots, \lambda_m$ and the monodromy group.

This problem is reduced to the Riemann boundary problem. Draw the cuts $\Gamma_j, (\lambda_0, \lambda_j), \Gamma = \bigcup_{j=1}^{m} \Gamma_j$, where λ_0 is an arbitrary point (see Figure 9).[2] We have the boundary problem

$$\Psi_+(\lambda) = G(\lambda)\psi_-(\lambda), \qquad G(\lambda) = G_j \quad \text{as} \quad \lambda \in \Gamma_j. \tag{24.27}$$

[2]In Figure 9 $\lambda_0, \ldots, \lambda_m, \Gamma_0$ are denoted by $P_a, \ldots, P_m, L$, respectively. (See page 37).

Note that the contour Γ has the self-intersection at the point λ_0, and the function $G(\lambda)$ has a jump at this point.

Therefore, there arises the necessity to study matrix Riemann boundary problems with discontinuous coefficients and contours with self-intersections. Such problems were studied by Röhrl [c].

Here we shall treat another approach to the Riemann–Hilbert problem due to Röhrl [d] Arnol'd & Il'iašenko [2]. Draw the contour Γ_0 as in Figure 9, such that the λ-plane will be separated into the domain $T^+ \ni \lambda_1, \ldots, \lambda_m$, and the domain $T^- \ni \infty$. Cover the domain T^+ by the simply connected domains $U_j (j = 1, \ldots, m)$, such that $\lambda_j \in U_j, U_j$ contains no other points λ_k. In every domain U_j, we represent some matrix solution of Eq. (24.25) $\psi_+(\lambda)$ by

$$\begin{aligned} \psi_+(\lambda) &= (\lambda - \lambda_j)^{C_j} \Psi_j(\lambda), \qquad \lambda \in U_j, \\ G_j &= \exp 2\pi i C_j, \qquad j = 1, \ldots, m, \end{aligned} \tag{24.28}$$

where $\Psi_j(\lambda)$ are nondegenerate holomorphic matrices in U_j. The coboundary of the cochain $\{\Psi_j\}$ is

$$\Psi_{jk} = \Psi_j \Psi_k^{-1} = (\lambda - \lambda_j)^{C_j} (\lambda - \lambda_k)^{-C_k}, \qquad \lambda \in U_j \cap U_k. \tag{24.29}$$

Therefore, we define the vector bundle with transition matrices $\{\Psi_j k\}$ over T^+. As long as T^+ is a Stein manifold, this bundle is trivial. (Cartan [a]), and there exist holomorphic nondegenerate matrices H_j in U_j, such that $\Psi_{jk} = H_j^{-1} H_k$, and hence

$$X_+(\lambda) = H_j(\lambda)\Psi_j(\lambda) = H_k(\lambda)\Psi_k(\lambda)$$

is a holomorphic matrix in T^+, and

$$\begin{aligned} \psi_+(\lambda) &= (\lambda - \lambda_j)^{C_j} H_j^{-1}(\lambda) X_+(\lambda) \\ &= (\lambda - \lambda_j)^{C_j} F_j(\lambda), \\ F_j(\lambda) &= H_j^{-1}(\lambda) X_+(\lambda), \qquad \lambda \in U_j. \end{aligned} \tag{24.30}$$

The matrix $\psi_+(\lambda)$ satisfies the equation

$$\psi_+^{-1}(\lambda) \frac{d\psi_+}{d\lambda} = F_j^{-1} \left(\frac{dF_j}{d\lambda} + \frac{C_j}{\lambda - \lambda_j} F_j \right). \tag{24.31}$$

On the λ-plane the fundamental matrix has the form

$$\begin{aligned} \psi(\lambda) &= \psi_+(\lambda) X_+(\lambda), \qquad \lambda \in T^+, \\ \psi(\lambda) &= \left(\frac{1}{\lambda}\right)^{C_{m+1}} X_-(\lambda), \qquad \lambda \in T^-, \\ \exp 2\pi i C_{m+1} &= (G_1, \ldots, G_m)^{-1}. \end{aligned} \tag{24.32}$$

If the point λ moves along Γ_0 encircling infinity, then we see that it passes around all points $\lambda_1, \ldots, \lambda_m$ in the negative direction. The matrices $X_\pm(\lambda)$ are supposed to be nondegenerate and holomorphic in $T^\pm$. We obtain the problem

$$X_+(\lambda) = \psi_+^{-1}(\lambda) \left(\frac{1}{\lambda}\right)^{C_{m+1}} X_-(\lambda), \qquad \lambda \in \Gamma_0. \tag{24.33}$$

In general, this problem has a solution with a pole or zero at infinity. In this case, the matrix $\psi^{-1}\frac{d\psi}{d\lambda}$ may have a pole at infinity whose order is more than 1.

Indeed, in the domain T^{-}, the matrix $X_{-}(\lambda)$ is represented in the form (Sonvage's lemma, [19])

$$\psi(\lambda) = \lambda^{D} F(\lambda) P(\lambda), \tag{24.34}$$

where D is a constant diagonal matrix and $P(\lambda)$ is polynomial. The matrix $\tilde{\psi}(\lambda) = \psi(\lambda) P^{-1}(\lambda)$ has only Fuchsian singularities at the points $\lambda_1, \dots, \lambda_m$. If C_{m+1} is diagonal, then

$$\tilde{\psi}(\lambda) = \left(\frac{1}{\lambda}\right)^{C_{m+1}-D} F(\lambda), \tag{24.35}$$

and hence infinity is a Fuchsian singularity, i.e., $\tilde{\psi}^{-1}(\tau)\frac{d\tilde{\psi}}{d\tau}$, $\tau = \lambda^{-1}$ has the pole of the first order at infinity.

As long as all points $\lambda_1, \dots, \lambda_m$ are equal, it is sufficient that any of the matrices $G_1, \dots, G_{m+1}$ can be reduced to the diagonal form.

In the opposite case, the Riemann–Hilbert problem may be unsolvable (to the contrary of widespread opinion). For example, if

$$D = \begin{pmatrix} 0 & 1 \\ 0 & 0, \end{pmatrix} \qquad C = \begin{pmatrix} 0 & 1 \\ 0 & 0 \end{pmatrix}, \qquad \tau = \lambda^{-1}, \qquad \psi = \lambda^{D}\lambda^{-C},$$

then

$$\frac{d\psi}{d\lambda}\psi^{-1} = \begin{pmatrix} 0 & -1/\tau^2 \\ 0 & 1\tau^2 \end{pmatrix}$$

(see Arnol'd & Il'iašenko [2]).

In conclusion, we point out one simple interpretation of this problem (Jimbo, Miwa & Sato [a]).

As is known from [39], creation operators $\psi^{(i)}(\lambda)$ and annihilation operators $\psi^{*(i)}(\lambda)$ of the free fermion field are connected by the commutation relations

$$\begin{gathered} \psi^{(i)}(\lambda)\psi^{*(j)}(\lambda') + \psi^{*(j)}(\lambda')\psi^{(i)}(\lambda) = \delta_{ij}\delta(\lambda - \lambda'), \\ \psi^{(i)}(\lambda)\psi^{(j)}(\lambda') + \psi^{(j)}(\lambda')\psi^{(i)}(\lambda) = 0, \\ \psi^{*(i)}(\lambda)\psi^{*(j)}(\lambda') + \psi^{*(j)}(\lambda')\psi^{*(i)}(\lambda) = 0. \end{gathered} \tag{24.36}$$

Let the values λ_j be real and $\lambda_1 < \cdots < \lambda_m$. Define the matrices

$$\begin{aligned} (m_{ik}) &= G_j \dots G_n, & \lambda_{j=1} < \lambda < \lambda_j, \\ (m^*_{ik}) &= {}^t(G_j \dots G_n)^{-1}, & j = 1, \dots, m. \end{aligned} \tag{24.37}$$

Here ${}^tA$ is the matrix transposed to a matrix A. If φ is the field operator, such that

$$\begin{aligned} \varphi\psi^{(j)}(\lambda) &= \sum \psi^{(i)}\varphi m_{ij}(\lambda), \\ \varphi\psi^{*(j)}(\lambda) &= \sum \psi^{*(i)}\varphi m^*_{ij}(\lambda), \end{aligned} \tag{24.38}$$

then can may show that the matrices

$$
\begin{aligned}
Y_+(\lambda) &= -2\pi i(\lambda_0 - \lambda)(< \psi^{*(i)}(\lambda_0)\psi^{(j)}(\lambda)\frac{\varphi}{<\varphi>} >), \\
Y_-(\lambda) &= -2\pi i(\lambda_0 - \lambda)(< \psi^{*(i)}(\lambda_0)\frac{\varphi}{<\varphi>}\psi^{(j)}(\lambda) >), \\
i, j &= 1, \dots, n
\end{aligned} \tag{24.39}
$$

are analytic in the domains $T^{\pm}$, respectively, and satisfy the boundary condition

$$
\begin{aligned}
Y_-(\lambda) &= Y_+(\lambda)T(\lambda), \\
T(\lambda) &= G_j, \dots, G_m, \qquad \lambda_{j-1} < \lambda < \lambda_j, \\
j &= 1, \dots, m.
\end{aligned} \tag{24.40}
$$

APPENDIX 1

HYPERELLIPTIC SURFACES

Consider the hyperelliptic surface M defined by the equation

$$\text{(A1.1)} \qquad w^2 = R_{2g+1}(z), \qquad R_{2g+1}(z) = \prod_{j=0}^{2n} (z - z_j),$$

with the branch points $z_0, \ldots, z_{2n}$ and ∞ (see Figure 3, and [31]). The genus of the surface M is equal to g.

Abelian differentials of the first kind are

$$\text{(A1.2)} \qquad \omega_j(p) = \frac{z^j\, dz}{\sqrt{R_{2g+1}(z)}}, \qquad j = 0, \ldots, g-1, \qquad p = (z, \pm).$$

It is necessary to verify that $\omega_j(p)$ are regular at the branch points and at infinity.

At infinity we use the local coordinate $t = (\sqrt{z})^{-1}$. We have

$$\omega_j(p) = \frac{t^{2g+1}\, dt}{t^{2j+2}\sqrt{\widetilde{R}(t)}} = \frac{t^{2g-2j-1}\, dt}{\sqrt{\widetilde{R}(t)}},$$

$$\widetilde{R}(t) = \prod_{j=0}^{2g} (1 - z_j t^2), \qquad \widetilde{R}(0) \neq 0,$$

$$j = 0, \ldots, g-1.$$

At the points $z = z_k$, we use the local coordinate $t = \sqrt{z - z_k}$. Then

$$R_{2g+1}(z) = (z - z_k) r_k(z) = t^2 \tilde{r}_k(t),$$

$$\tilde{r}_k(t) = \prod_{j \neq k}^{2g} (t^2 + z_k - z_j), \qquad \tilde{r}_k(0) \neq 0,$$

$$\omega_j(p) = \frac{2(t^2 + z_k)^j\, dt}{\sqrt{\tilde{r}_k(t)}}.$$

Therefore, an arbitrary Abelian differential of the first kind on the hyperelliptic surface M has the form

$$\omega(p) = \frac{P_{g-1}(z)dz}{\sqrt{R_{2g+1}(z)}}, \tag{A1.3}$$

where $P_{g-1}(z)$ is an arbitrary polynomial of the degree $g-1$ with complex coefficients. Whence it follows that any Abelian differential of the first kind has $g-1$ zeros on every sheet of the surface M.

Now, let the divisor $\gamma = \sum_1^g p_j$ have the degree g, and projections of all the points p_j on the z-plane are different. Since ω has only $g-1$ different zeros on the z-plane, $\dim H(\gamma) = 0$. By the Riemann–Roch theorem,

$$\dim L(\gamma) = \dim H(\gamma) + \deg \gamma - g + 1 = 1.$$

This means that the function

$$\theta(w_s(p) - \sum_{j=1}^{g} w_s(p) - k_s) \tag{A1.4}$$

is nonzero.

We deduce the formulae for the Abelian differentials of the first and third kind. The Abelian differential of the second kind with the pole of the second order at the point $z(p) = a \neq \infty$ has the form

$$d\tilde{t}_p = \frac{1}{2(z-a)^2\sqrt{R_{2g+1}(z)}}\Big[\sqrt{R_{2g+1}(z)} + \sqrt{R_{2g+1}(a)} + \left(\sqrt{R_{2g+1}(z)}\right)'(z-a)\Big]dz. \tag{A1.5}$$

If $z(a) = \infty$, then

$$d\tilde{t}_p = \frac{z^g dz}{2\sqrt{R_{2g+1}(z)}}. \tag{A1.6}$$

The Abelian differential of the third kind with poles $z(p_1) = a_1, z(p_2) = a_2$ has the form

$$d\tilde{\omega}_{p_1p_2} = \frac{1}{2\sqrt{R_{2g+1}(z)}} \left[\frac{\sqrt{R_{2g+1}(z)} + \sqrt{R_{2g+1}(a_2)}}{z-a_2} - \frac{\sqrt{R_{2g+1}(z)} + \sqrt{R_{2g+1}(a_1)}}{z-a_1}\right] dz. \tag{A1.7}$$

The normalized Abelian differentials dt_p^1 and $d\omega_{p_1p_2}$ may be obtained by addition of corresponding combinations of Abelian differentials of the first kind.

APPENDIX 2

THE MATRIX RIEMANN PROBLEM ON THE PLANE

Consider the matrix Riemann problem

$$\Phi_+(t) = G(t)\Phi_-(t), \qquad t \in L \tag{A2.1}$$

on the complex plane. Here $G(t)$ is a nondegenerate $n \times n$-matrix satisfying the Hölder condition, $\Phi_\pm(z)$ are vectors or $n \times n$-matrices analytic in the domains $T^\pm$, respectively [27]. The values $\Phi_\pm(z)$ are supposed to have a finite number of poles.

A vector solution having a pole of the order m at infinity is represented by

$$\Phi_\pm(z) = \frac{1}{2\pi i}\int_L \varphi(\tau)\frac{d\tau}{\tau - z} + P_m(z), \tag{A2.2}$$

where $P_m(z)$ is a polynomial of the degree m with vector coefficients. Substituting (A2.2) in (A2.1), we get the singular integral equation system

$$\frac{1+G(t)}{2}\varphi(t) + \frac{1-G(t)}{2\pi i}\int_L \varphi(\tau)\frac{d\tau}{\tau - t} = [1 - G(t)]P_m(t) \tag{A2.3}$$

of the index

$$\kappa = \operatorname{ind}_L \det G.$$

The number l of solutions of Eq. (A2.3) is

$$l \geqslant \kappa + (m+1)n - l', \tag{2.4}$$

where $(m+1)n$ is the number of free coefficients of the right-hand side of (A2.3), and l' is the number of solutions of the adjoint system (and hence the number of solvability conditions). Therefore, if m is sufficiently large, there exist n vector solutions of (A2.1) forming the $n \times n$ matrix $F(z)$, such that $\det F(z) \not\equiv 0$.

Note that the matrix solution $F(z)$ satisfies the condition

$$\det F_+(t) = \det G \det F_-(t), \qquad t \in L, \tag{2.5}$$

and hence the function $\det F(z)$ has $N + \kappa$ zeros (N is the order of the pole at infinity). Using appropriate linear combinations of the columns of $F(z)$ with rational coefficients, it is possible to get the matrix solution $X(z)$ which is called canonical and possesses the following two properties. The matrix $X(z)$ is nondegenerate for any finite z, and the order κ of zero of $X(z)$ at infinity (a pole of order $|\kappa|$ if $\kappa < 0$) is equal to the sum of orders κ_j of the columns $X_j(z), j = 1, \ldots, n$. The integers κ_j are called the partial indices of the problem and are independent of a chosen canonical solution.

If the point $z = 0$ belongs to T^+, the matrix $G(t)$ may be factored in the form

$$G(t) = H_+(t)\,\mathrm{diag}(t^{\kappa_1}, \ldots, t^{\kappa_n})H-(t), \tag{A2.6}$$

where $H_\pm(z)$ are nondegenerate matrices holomorphic in $T_\pm$. The representation (A2.6) is unique.

We deduce two criteria for $\kappa_1 = \cdots = \kappa_n = 0$, if $\kappa = 0$.

Criterion 1. So that problem (A2.1) possesses zero partial indices, it is necessary and sufficient that the problem have a matrix holomorphic solution.

Criterion 2. (Gohberg & Krein [14])

For all partial indices of the problem (A2.1) to be zeros, it is sufficient that one of the matrices $G(t) + G^+(t)$ or $i[G(t) - G^+(t)]$ be definite.[1]

For the following results in this area, see, for example, Krein & Spitkovskii [a], Spitkovskii [a,b,c], Bart, Gohberg & Kaashoek [a,b,c,d,e]. The partial indices κ_j are nonstable (see Gohberg & Krein [14], Grothendieck [a], Bojarsky [a,b]). These integers define the vector bundle B_G over the sphere corresponding to the problem (A2.1) (Grothendieck [a]).

[1] The matrix (a_{ij}) is definite if the quadratic form $\sum_{i,j} a_{ij}x_i x_j$ takes the values of the same sign.

APPENDIX 3

ONE APPROXIMATE METHOD OF SOLVING THE MATRIX RIEMANN PROBLEM

Here we deduce one approximate method of solving of the matrix Riemann problem.

Let

$$G(p) = R_+^{-1}(p)[1 + g(p)]R_-(p), \qquad p \in L, \tag{A3.1}$$

where matrices $R_\pm(q)$ are analytic in the domains $T^\pm \subset M$, respectively, and have a finite number of poles and zeros, 1 is the unit matrix, and the matrix $g(p) = (g_{ij}), |g_{ij}| < g_0$ and satisfies the Hölder condition. The constant g_0 will be fixed below.[2]

Consider the problem

$$\Phi_+(p) = G(p)\Phi_-(p). \tag{A3.2}$$

Assume

$$F_\pm(q) = R_\pm(q)|F_\pm(q), \qquad q \in T^\pm. \tag{A3.3}$$

We get the problem

$$F_+(p) - F_-(p) = g(p)F_-(p). \tag{A3.4}$$

The problem (A3.4) will be studied in the Hardy classes $H_2(T^\pm)$ defined below. Let $d\alpha(p)$ be some real differential on the contour L. The space $L_2^n(L)$ (the

[2]Let $G() = H(p) + h(p)$, where $H(p) = R_+^{-1}(p)R_-(p)$. Then $G(p)$ has the form (A3.1) where $g(p) = R_+(p)h(p)R_-^{-1}(p)$. If matrix elements of $h(p)$ are small, the matrix elements of $g(p)$ are also small.

index n may be omitted) is the Hilbert space of the vector functions $F(p) = (F_1, \ldots, F_n)$ with the scalar product

$$(F, F') = \sum_{i=1}^{n} \int_L F_j(p)\overline{F'_j(p)}d\alpha(p). \tag{A3.5}$$

Let δ be a divisor such that $\deg \delta = g$ and $\dim \delta = 1$. For simplicity, we suppose that $\delta \in T^-$. We also fix an arbitrary point $q_0 \in T^-, q_0 \overline{\in} \delta$. The Hardy classes $H_2(T^{\pm})$ (more exactly, $H_2(T^{\pm}, \delta, q_0, d\alpha, n)$) are the classes of vector functions which are analytic in $T^{\pm}$, respectively, and are multiples of the divisor $-\delta + q_0 \in T^+$. The boundary values of these functions have to belong to the space $L_2^n(L)$. We use two properties of these classes (see, for example, [8])

a) If vector-functions $F_{\pm}(q) \in H_2(T^{\pm})$, and

$$F_+(p) = F_-(p), \qquad p \in L,$$

then $F_{\pm}(q)$ form single analytic vector functions on M (which are multiples of $q_0 - \delta$).

b) Vector functions belonging to the classes $H_2(T^{\pm})$ are represented by the Cauchy integrals. The Plemelj–Sokhotsky formulae (4.12) are valid for a Cauchy type integral with the density of the class $L_2(L)$.

We search for a solution of the problem (A3.4) of the form

$$F_{\pm}(q) = 1 + \frac{1}{2\pi i}\int_L \varphi(p)M(p,q)dz(p) \tag{A3.6}$$

with the characteristic divisor δ and the polar point q_0. We get the singular equation systems

$$\varphi(q) = g(q)\left[-\frac{1}{2}\varphi(q) + \frac{1}{2\pi i}\int_L \varphi(p)M(p,q)dz(p)\right] + g(q) \tag{A3.7}$$

in the space $L_2(L)$. The operator

$$A\varphi = -\frac{1}{2}\varphi(q) + \frac{1}{2\pi i}\int_L \varphi(p)M(p,q)dz(p), \qquad q \in L \tag{A3.8}$$

is a projector from the space $L_2(L)$ on the space ImA of boundary values of vector functions of the class $H_2(T^-)$. The kernel Ker A of the operator A is the space of boundary values of vector functions of the class $H_2(T^+)$.

Let $\varphi(q) \in L_2(L)$. Letting

$$F(q) = \frac{1}{2\pi i}\int_L \varphi(p)M(p,q)dz(p),$$

we conclude that

$$\varphi(p) = F_+(p) - F_-(p), \qquad p \in L, \qquad F_\pm(q) \in H_2(T^\pm). \tag{A3.9}$$

Therefore

$$L_2(L) = \operatorname{Ker} A \bigoplus \operatorname{Im} A.$$

The representation (A3.9) is unique, since $\operatorname{Ker} A \cap \operatorname{Im} A = 0$. Indeed, if $f(q) \in \operatorname{Ker} A \cap \operatorname{Im} A$, then $f(q)$ is analytic on M and a multiple of the divisor $q_0 - \delta$. As long as $\dim \delta = 1$, $f(q) \equiv \text{const}$, and hence $f(q) \equiv 0$. As follows from the Riez theorem [8], the operator A is bounded in $L_2(L)$, $\|A\|_{L_2(L)} < \infty$. But, in general, the operator A is not an orthogonal projector, and hence $A_{L_2(L)} \geqslant 1$. The operator gA norm is $\leqslant g_0 \|A\|_{L_2(L)}$ (see (A3.1)). Therefore, if

$$g_0 < \frac{1}{\|A\|_{L_2(L)}}, \tag{A3.10}$$

the norm $\|gA\| < 1$, and hence iterations

$$\varphi_m(q) = g(q)\left[-\frac{1}{2}\varphi_{m-1}(q) + \frac{1}{2\pi i}\int_L \varphi_{m-1}(p) M(p,q)\, dz(p)\right] + g(q) \tag{A3.11}$$

converge in the space $L_2(L)$.

Let M be the hyperelliptic surface with the real branch points $z_0, \dots, z_{2g}$ and L be the set of cuts $(z_0, z_1), (z_2, z_3), \dots, (z_{2g}, \infty)$ (Figure 3). The Abelian differentials of the first kind are defined by Eq. (A1.5). Let $P_1(\xi_1, -), \dots, p_g(\xi_g, -)$ be g points belonging to T^- (ξ_j is a complex number, the signs $\pm$ define the sheet). As follows from Eq. (A1.5), there is no Abelian differential of the first kind which is a multiple of the divisor $\delta = \sum_1^g p_j$ (see Appendix 1). It means that there exists an Abelian differential $d\alpha(q)$ of the third genus with poles at the points $(\xi_0, -), (\bar{\xi}_0, +)$ and zeros $p_j (j = 1, \dots, g)$ which is real on L. Using the kernel $M(p, q)$ with the characteristic divisor $\delta = \sum_{j=1}^g p_j$ and the pole at the point $q_0(\xi_0, -)$, we conclude that in that case, $\operatorname{Ker} A \perp \operatorname{Im} A$. Indeed, let $\varphi(q) \in \operatorname{Im} A$ and $\psi(q) \in \operatorname{Ker} A$. Introduce the involution on M by the rule $q(\xi, +) \to \tilde{q}(\bar{\xi}, -)$. Then $\overline{\psi(\tilde{q})} \in H_2(T^-)$, and

$$\int_L \varphi(p)\overline{\psi(p)}\, d\alpha(p) = 0,$$

since the differential $\varphi(q)\overline{\psi(\tilde{q})}\, d\alpha(q)$ is holomorphic in the domain T^-. Therefore, A is an orthogonal projector, $\|A\| = 1$, and one may suppose $g_0 < 1$.

If the genus of the surface M is equal to zero, the iteration process generates a holomorphic solution, since $\delta = 0$. This means that all partial indices of the problem (A3.2) are equal to zero. In the general case, we get the solution which is a multiple of the divisor δ.

If the proposed method is not applicable, the Riemann problem is reduced to the integral equation, which may be solved by standard methods.

APPENDIX 4

THE RIEMANN–HILBERT BOUNDARY PROBLEM

Let T be a finite-connected domain on a Riemann surface bounded by the Liapounov contour L consisting of $m+1$ connected components, $L = \sum_{j=0}^{m} L_j$, and let $\lambda(p) = \alpha(p) + i\beta(p), p \in L, |\lambda(p)| = 1$, be a Hölder function on L.

The boundary Riemann–Hilbert problem determines a function $F(q)$ holomorphic in T, continuous up to the boundary L, and satisfying the boundary condition

$$\text{Re}[\overline{\lambda(p)}F(p)] = 0. \tag{A4.1}$$

For plane domains, this problem has wide applications. It was solved by Muskhelishvili [27], who reduced this problem to the Riemann boundary problem (for the case $m = 0$). The case of multiconnected domains is more complicated. For such domains, there is the so-called singular case [13]. We use the Muskhelishvili method for arbitrary finite domains (see Rodin [a], [b]). This provides a general approach explaining all singularities of the problem.

Let $\{U, z(U)\}$ be the complex structure of the surface T consisting of coordinate domains and local coordinate functions. Consider the surface $\tilde{T}$ which is obtained from T by the change of the structure $\{U, z(U)\}$ to the complex-conjugate structure $\{U, \overline{z(U)}\}$. For example, if T is a disk $|z| < 1$ or a halfplane $\text{Im}\, z > 0$, then T is a disk $|z| > 1$ or a half-plane $\text{Im} z < 0$, respectively. Then $M = T \cup \tilde{T} \cup L$ is a compact Riemann surface of a genus $g = 2g' + m$, where g' is a genus of T. This surface is called the double of T. For example, the double of the ring $r < |z| < R$ is a torus. There exists the natural involution: if $p \in T$, then $\tilde{p} \in \tilde{T}$ is the corresponding point with a complex conjugate local coordinate (for the disk $|z| < 1$ $\tilde{z} = \frac{1}{\bar{z}}$, for the halfplane $\text{Im} z > 0$ $\tilde{z} = \bar{z}$).

Let $F(q)$ be a solution of the problem (A4.1). Assume

$$\begin{aligned} F_+(q) &= F(q), \qquad q \in T = T^+, \\ F_-(q) &= \overline{F(\tilde{q})}, \qquad q \in \tilde{T} = T^-, \qquad \tilde{q} \in T^+. \end{aligned} \tag{A4.2}$$

Then the boundary condition (A4.1) may be rewritten in the Riemann problem form

$$\overline{\lambda(p)}F_+(p) + \lambda(p)F_-(p) = 0, \qquad p \in L \tag{A4.3}$$

of the index $2\kappa, \kappa = \mathrm{ind}_L \lambda(p)$ for double M. This explains the singular case, since a multiconnected plane domain T generates the Riemann problem on a surface of a nonzero genus.

Conversely, every solution of the problem (A4.3) possessing the symmetry (A4.2) is a solution of the problem (A4.1). If $\Phi_{\pm}(q)$ is an arbitrary solution of the problem (A4.3), then the solution

$$F_{\pm}(q) = \frac{1}{2}[\Phi_{\pm}(q) + \overline{\Phi_{\mp}(\tilde{q})}] \tag{A 4.4}$$

is symmetric. This means that if l is the complex dimension of the solutions space of the problem (A4.3) then $2l$ is a real dimension of the solution space for the problem (A4.1).

The adjoint problem to (A4.3) is

$$\Psi_{+}(p)dz(p) = -\frac{\overline{\lambda(p)}}{\lambda(p)}\Psi_{-}(p)dz(p). \tag{A4.5}$$

Let $S(p)$ be the real positive parameter on the contour L (for example, the arc length). Then

$$\lambda(p)\Psi_{+}(p)\frac{dz(p)}{ds} + \overline{\lambda(p)}\Psi_{-}(p)\frac{dz(p)}{ds} = 0. \tag{A4.6}$$

If the solution of (A4.5) has the symmetry

$$\Psi(\tilde{q})dz(\tilde{q}) = \overline{\Psi(q)dz(q)}, \tag{A4.7}$$

then we obtain the boundary problem for differentials

$$\mathrm{Re}\left[\lambda(p)\Psi(p)\frac{dz(p)}{ds}\right] = 0, \qquad s \in L \tag{A4.8}$$

adjoint to (A4.1). If $2h$ is a real dimension of the solution space of (A4.8), then

$$l - h = 2\kappa - 2g' - m + 1. \tag{A4.9}$$

For solvability of the nonhomogeneous problem

$$\mathrm{Re}\left[\overline{\lambda(p)}F(p)\right] = g(p), \tag{A4.10}$$

it is necessary and sufficient that

$$\int_L g(p)\lambda(p)\Psi_j(p)dz(p) = 0, \qquad j = 1, \dots, 2h, \tag{A4.11}$$

where $\Psi_j(p)dz(p), j = 1, \dots, 2h$ is a complete system of solutions of the adjoint problem (A4.8).

NOTATIONS

Abelian differentials

of the first kind

$dw_j, dW_j, d\theta_j,$ pp. 9, 10, 127
$\omega_j(p)$ p. 179

of the second kind

$dt^n_{p,z}, dt_p, dT^n_{p,z}, dT_p$ p. 10
$d\tilde{t}_p$ p. 180

of the third kind

$d\omega_{p_0p}, d\Omega_{p_0p}$ pp. 11, 127
$d\tilde{\omega}_{p_0p}$ p. 180

Differential spaces

$\Gamma_c(D), \Gamma_e(D), H(D)$ pp. 6, 7, 101, 102
$\Gamma_{eo}, \tilde{\Gamma}_{eo}, \Gamma_h$ p. 102
$\Gamma(T, A^{\alpha,\beta})$ — space of the forms of the type (α, β) in the domain T p. 56
$\Gamma(F), \Gamma(T, F)$ — space of the sections of a sheaf F (in the domain T) p. 56
Operator * pp. 6, 101
$\omega_1 \wedge \omega_2$ — outer product p. 6

Cauchy kernels

$M^*(p,q)dz(p), M_*(p,q)dz(p)$ pp. 20, 128
$M(p,z)dz(p)$ pp. 23, 170
δ — characteristic divisor p. 24
q_0 — polar point p. 24

Groups

$Cn^q(F), Cn^q_D(F)$ — cochain groups pp. 55, 103
$Z^q(F), Z^q_D$ — cocycle groups pp. 55, 103
$H^q(F), H^q_D(F)$ — cohomology groups pp. 56, 103
$\xi(F)$ — Euler characteristic p. 63

δ, δ^* — coboundary homomorphism p. 56
$\Gamma(F) = H^0(F)$ p. 56

Sheaves

$A^{\alpha,\beta}, C^{\alpha,\beta}, Q^1, Q, Q^*, M^*$ pp. 55, 59
$A^{\alpha,\beta}(B), C^{\alpha,\beta}(B), Q^1(B), Q(B), Q^*(B)$ p. 61
$Q^{\alpha,\beta}(B)$ — sheaf of germs of holomorphic sections-forms of type (α,β) of a bundle B p. 61
$C_z^{\alpha,\beta}(B)$ — sheaf of gerrms of holomorphic forms of the type (α,β) p. 61
$Q(B) = Q^{0,0}(B)$ — sheaf of germs of holomorphic sections of a bundle B p. 61
$Q_\gamma = Q(-B_\gamma)$ — sheaf of germs of analytic functions which are multiples of the divisor γ p. 62
$Q_G = Q(B_G)$ pp. 69, 84, 117

Bundles

B_G — bundle determined by the Riemann boundary problem pp. 69, 83, 117
B_h — bundles determined by the divisor h p. 61
K — tangential bundle p. 62

Other notations and terms

$M_n \searrow M$ — normal exhaustion p. 96
$\lambda(L)$ — extremal length p. 99
$D(\omega)$ — Dirichlet integral of ω p. 103
D — norms p. 103
D-divisors p. 106
S-divisors p. 111
S-boundary problem p. 124
J, J_D, J_S — Jacobi varieties pp. 41, 123
$S_0, S_1 = H_D^1(Q)/S_0$ — singular group pp. 105, 118
$\deg(\gamma)$ — divisor degree pp. 17, 107
$L(\gamma)$ — space of functions which are multiples of the divisor $-\gamma$
$\dim \gamma = \dim L(\gamma)$ p. 18
$H(\gamma)$ space of differentials which are multiples of the divisor γ p. 18
$\dim H(\gamma) = \dim(K - \gamma)$ p. 18
κ — Cauchy index of a Riemann boundary problem pp. 28, 117
Cochains different from zero on ideal boundary p. 121
Cochains bounded on ideal boundary p. 121
Solution of a Riemann problem p. 117
Strong, weak solutions of a Riemann problem p. 122
$\{U_i, i \in I\}$ surface covering p. 102
$\{U_i, i \in I_0\}$ surface triangulation p. 103

N — covering constant p. 102
Special covering p. 108
$\gamma(p)$ analytic function in some domain determined
by the given divisor, $(\gamma(p)) = \gamma$ p. 52
(f) — divisor of zeros and poles of the function $f(p)$ p. 17

References

Monographs and Expository Papers

1. Ahlfors, L. V., Sario, L., *Riemann Surfaces*, Princeton Univ. Press, Princeton, 1960.
2. Arnol'd, V. I., Il'iašenko, Yu. S., "Ordinary differential equations," (Russian), in *Dynamic Systems I*, VINITI, Moscow, 1986.
3. Bers, L., *Pseudoanalytic Functions*, New York, 1953.
4. Bers, L., John F., Schechter, M., *Differential Equations*, Intern. Publ., New York, 1964.
5. Cartan, H., Eilenberg, S., *Homological Algebra*, Princeton University Press, Princeton, 1957.
6. Chadan, K., Sabatier, P. C., *Inverse Problems in Quantum Scattering Theory*, Springer, New York, 1977.
7. Chern, S. S., *Complex Manifolds*, Univ. of Chicago Press, Chicago, 1956.
8. Danilük, I. I., *Nonregular Boundary Problems on the Plane* (Russian), Nauka, Moscow, 1975.
9. Dubrovin, B. A., Matveev, V. B., Novikov, S. P., "Nonlinear equations of Korteveg-de Vries type, finite-gap linear operators and Abelian varieties," Uspehi Matem. Nauk **31** (1976), p. 107.
10. Fay, H., *Theta Functions on Riemann Surfaces*, Lecture Notes in Mathematics **352**, Springer, New York, 1973.
11. Fokas, A. S., Ablowitz, M. J., *Nonlinear Phenomena*, Lecture Notes in Physics **189**, Springer, New York, 1982.
12. Forster, O., *Riemannsche Flächen*, Springer, Berlin, Heidelberg, New York, 1977.
13. Gahov, F. D., *Boundary Value Problems*, Pergamon, New York, 1966.
14. Gohberg, I. C., Krein, M. G., "Systems of integral equations on semiaxis with kernel depending on the arguments difference," Uspehi Matem. Nauk **13** (1958), 3–72.
15. Griffiths, P., Harris, J., *Principles of Algebraic Geometry*, J. Wiley, New York, 1978.

16. Gunning, R. C., *Lectures on Riemann Surfaces*, Princeton Math. Notes **2** (1966).
17. Gunning, R. C., *Lectures on Vector Bundles over Riemann Surfaces*, Princeton Math. Notes **6** (1967).
18. Gunning, R. C., *Lectures on Riemann Surfaces: Jacobi Varieties*, Princeton Math. Notes **12** (1972).
19. Hartman, P., *Ordinary Differential Euqations*, J. Wiley, New York, 1964.
20. Hirzebruch, F., *Topological Methods in Algebraic Geometry*, Springer, New York, 1966.
21. Hörmander, L., *An Introduction to Complex Analysis in Several Variables*, North-Holland, Amsterdam, 1973.
22. Krasnoselski, M. A., *Topological Methods in the Theory of Nonlinear Integral Equations*, Macmillan, New York, 1964.
23. Kosevich, A. M., Ivanov, B. A., Kovalev, A. S., *Magnetization Nonlinear Waves. Dynamical and Topological Solitons* (Russian), Naukova Dumka, Kiev, 1984.
24. Lamb, G. L., *Elements of Soliton Theory*, J. Wiley, New York, 1980.
25. Levitan, B. M., *Inverse Sturm-Liouville Problems* (Russian), Nauka, Moscow, 1984.
26. Levitan, B. M., Sargsjan, I. S., *Introduction to spectral theory*, Transl. Math. Monogr. Amer. Math. Soc. **39** (1975), Providence, RI.
27. Muskhelishvili, N. I., *Singular Integral Equations*, Noordhoff, Groningen, 1953.
28. Nevanlinna, R., *Uniformisierung*, Springer, Berlin, 1953.
29. Rodin, Yu. L., *Lectures on Riemann Surfaces*, Parts 1, 2; Institute of Solid State Physics of he Acad. of Sciences of the USSR, Chernogolovka, 1975; 1976 (Russian).
30. Rodin, Yu. L., *Generalized Analytic Functions on Riemann Surfaces*, Lecture Notes in Mathematics, Springer, **7** 1987. (To be published).
31. Springer, G., *Introduction to Riemann Surfaces*, Addison-Wesley, Reading, MA, 1957.
32. Shiffer, M., Spencer, D. C., *Functionals of Finite Riemann Surfaces*, Princeton Univ. Press, Princeton, 1954.
33. Titchmarsh, E. C., *Eigenfunction Expansions Associated with Second-Order Differential Euqations*, Clarendon Press, Oxford, 1946.
34. Tyurin, A. I., "Classification of vector bundles over algebraic curves," Izvestia Akad. Nauk, USSR, S. Math. **29** (1965), 658–680.
35. Vekua, I. N., "Systems of first-order differential equations and boundary problems with applications to the shell theory," Matem. sbornik **31** (1952), 217–314.
36. Vekua, I. N., *Generalized Analytic Functions*, Pergamon Press, London; Addition-Wesley, Reading, 1962.
37. Wendland, W. L., *Elliptic Systems in the Plane*, Pitman, London, 1979.
38. Weyl, H., *Die Idee der Riemannschen Fläche*, 3 Angl., Teubner, Stuttgart, 1955.
39. Zaiman, J. M., *Elements of Advanced Quantum Theory*, Cambridge Univ. Press, Cambridge, 1969.

40. Zakharov, V. E., Novikov, S. P., Manakov, S. V., Pitaevskii, L. P., *Solitons Theory. The Inverse Problem Method* (Russian), Nauka, Moscow, 1980.
41. Zverovich, E. I., "Boundary problems of analytical functions in Hölder classes on Riemann surfaces" (Russian), Uspehi Matem. Nauk **26** (1971), p. 113.
42. Rodin, Yu.L., "The Riemann boundary value problem on closed Riemann surfaces and integrable systems." Physica D, **24D** (1987), 1–53.

PAPERS

Abdulaev, R. N., [a] Sov. Math. Dokl. **4** (1963), p. 1525; [b] (Russian), Ucen. Zap. Perm. Gos. Univ. **103** (1963), p. 3; [c] (Russian), Bull. Acad. Sci. George. SSR **35** (1964), p. 512; [d] Sov. Math. Dokl. **6** (1965), p. 206; [e] (Russian), Ucen. Zap. Perm. Gos. Univ. **103** (1963), p. 143.

Ablowitz, M. J., Bar Yaacov, D., Fokas, A. S., [a] Stud. Appl. Math. **69** (1983), p. 135.

Ablowitz, M. J., Kaup, D. J., Newell, A. C., Segur, H., [a] Stud. Appl. Math. **53** (1974), p. 249.

Bart, H., Gohberg, I. C., Kaashoek, M. A., [a] "Operator theory," in *Advances and Appl.*, vol. 1, Birkhäuser, Basel, 1979; [b] *Ibid.*, **4** (1982); [c] Integral Equations and Operator Theory **5/3** (1982), p. 283; [d] Wiskundig Seminarium der Vrije Univ. Amsterdam **Rap. 231** (1983; [e] Proclamations M.I.N.S. (1983).

Beals, R., Coifman, R. R., [a] Comm. Pure Appl. Math. **37** (1984), p. 39; [b] "Multidimension scattering and inverse scattering" (preprint), Yale Univ. (1984).

Bers, L., [a] Proc. Nat. Acad. Sci. **37** (1951), p. 42; [b] Ann. Math. Study **30** (1953).

Bers L., Nirenberg, L., [a] Atti del. conv. intern. Salla Equazioni alle derivate parz. Trieste (1954), 111–140; [b] *Ibid.*, 141–167.

Behnke, H., Stein, K., [a] Math. Ann. **120** (1949), p. 430.

Bikbaev, R. F., Bobenko, A. I., [a] "On finite-gap integration of the Landau–Lifshitz equation X-Y-Z case" (preprint E-8-83), LOMI, Leningrad (1983).

Bikbaev, R. R., Bobenko, A. I., Its, A. R., [a] "The Landau–Lifschitz equation. The theory of explicit solutions," (Russian), (preprint DONFTI-84-6-7), Doneck (1982); [b] Dokl. Akad. Nauk USSR **272** (1983).

Bobenko, A. I., [a] (Russian), Zap. nauchn. seminar, LOMI **123** (1983), p. 58; [b] (Russian), Funct. Anal. Appl. **4** (1984), p. 15.

Bogdan, M. M., Covalev, A. C., [a] Pisma JETF **8** (1980), p. 453.

Bojarskii, B. V., [a] (Russian), Bull. Acad. Sci. Georg. SSR **21** (1958), p. 391; [b] *Modern Problems of Theory of Functions of Complex Variables*, (Russian), (1961), p. 57, Fizmatgiz, Moscow; [c] (Russian), Dokl. Akad. Nauk USSR **119** (1958).

Borisov, A. B., [a] Fiz. metal. i matalloved **55** (1983), p. 230.

Borovic, A. E., [a] Pisma JETF **28** (1978). p. 629.

Cartan, H., [a] "Colloque sur les Fonctions du Plusieurs Variables Tenu à Bruxelles," Liège-Paris (1953).
Chibrikova, L. I., [a] Izv. Visshih Ucebn. Zaved. Matematika **6** (1961), p. 121; "Letter to editor" (Russian), *Ibid.* **3** (1962).
Cotlazov, V. P., Khruslov, E. Ya., [a] Uspehi Matem. Nauk, **40** (1985) p. 197.
Dubrovin, B. A., [a] Funct. Anal. Appl. **9** (1975), p. 65; [b] Uspehi Matem. Nauk **31** (1976).
Ermakova, V. D., [a] Dokl. Acad. Nauk Ukr. SSR, S.A. (1982), p. 3–6.
Firsova, N. E., [a] Matem. Zametki **18** (1975), p. 831.
Flaschka, H., Newell, A., [a] Comm. Math. Phys. **76** (1980), p. 65.
Fokas, A. S., Ablowitz, M. J., [a] Study Appl. Math. **69**, (1983), p. 211.
Gardner, C. S., Green, J. M., Kruskal, M. D., Miura, R. M., [a] Phys. Rev. Lett. **19** (1967), p. 1095.
Gehtman, I. M., Stankevich, I. V., [a] (Russian), Diff. Equations **18** (1981), p. 2269.
Grothendieck, A., [a] Amer. J. Math. **79** (1957), p. 716.
Gusman, S. Y., Rodin, Yu. L., [a] (Russian), Sibirsk. Matem. Journ. **3** (1962), p. 527; MR **25**, p. 2195.
Its, A. R., Matveev, V. B., [a] Theor. Math. Phys. **23** (1975), p. 51.
Jimbo, M., Miwa, T., Sato, M., [a] "Monodromy preserving deformation of linear differential equations and quantum field theory," (RIMS preprint 246), 1978.
Khruslov, E. Ya., [a] Matem. Sbornik **99** (1976), p. 261.
Koppelman, W., [a] Comm. Pure Appl. Math. **12** (1959), p. 13; [b] J. Math. Mech. **10** (1961), p. 247; [c] Bull. Amer. Math. Soc. **67** (1961), p. 371.
Krein, M. G., Spitkovskii, I. M., [a] Analysis Math. **9** (1983), p. 23.
Krichever, I. M., Novikov, S. P., [a] Funct. Anal. Appl. **12** (1978), p. 41.
Kusunoki, Y., [a] Mat. Coll. Sci. Univ. Kyoto S. A1 **32** (1959), p. 235.
Lakshmanan, M., [a] Phys. Lett. **61A** (1977), p. 53.
Lax, P., [a] Lect. Appl. Math. AMS **15** (1974), 85– 96; [b] Comm. Pure Appl. Math. **28** (1975), p. 141.
Litvinchuk, G. S., Spitkovskii, I. M., [a] Math USSR Sbornik **45** (1983), p. 205.
McKean, H., Van Moerbeke, P., [a] Invent. Math. **30** (1975), p. 217.
Marchenko, V. A., [a] Matem. Sbornik **95** (1974), p. 331; [b] Dokl. Akad. Nauk USSR **217** (1974), p. 276.
Marchenko, V. A., Ostrovskii, I. V., [a] Matem. Sbornik **97** (1975), p. 540.
Meiman, N. N., [a] Jour. Math. Phys. **18** (1977).
Merzlyakova, G. D., Rodin, Yu. L., [a] (Russian), Ucen. Zap. Perm. Gos. Univ. **103** (1963), p. 463; MR 32, 1373.
Mikhailov, A. V., [a] Physica **3D** (1981), p. 73; [b] Phys. Lett. **92A** (1982), p. 51; [c] Pisma JETF **30** (1980), p. 183; [d] Phys. Lett. **92A** (1982), p. 51.
Novikov, S. P., [a] Funct. Anal. Appl. **8** (1974), p. 43.
Novikov, S. P., Dubrovin, B. A., [a] Uspehi Matem. Nauk **29** (1976); [b] JETF **67** (1974), p. 2131.

Rodin, Yu. L., [a] (Russian), Dokl. Akad. Nauk USSR **129** (1959), MR 22, 1660; [b] Soviet Math. Dokl. **1** (1960); [c] (Russian) "Modern investigations on the theory of functions of complex variables", FM, Moscow, (1960), p. 436; MR **22**, 1118; [d] (Russian), Ucen. Zap. Perm. Gos. Univ. **17** (1960), p. 83; MR **26**, p. 2604b; [e] (Russian), Ucen. Zap. Perm. Gos. Univ. **17** (1960), p. 79; MR **76**, p. 2604a; [f] "Certain problems in the mathematics and mechanics" (in honour of M. A. Lavrent'ev) (Russian), (1961), p. 224, SB Acad. Sci. of the USSR, Novosibirski; MR **42**, p. 523; [g] "Modern investigations on the theory of functions of complex variables" (Russian), (1961), p. 419, FM, Moscow; [h] Soviet Math. Dokl. **3** (1962); [i] Soviet Math. Dokl. **3** (1962), p. 177; [j] (Russian), Ucen. Zap. Perm. Gos. Univ. **22** (1962), p. 56; [k] (Russian), Dokl. Akad. Nauk, USSR **150** (1963), p. 1228, MR **27**, p. 5905; [l] (Russian), Ucen. Zap. Perm. Gos. Univ. **103** (1963), p. 64, MR **31**, p. 3601; [m] "The elliptic operators of the first order," (1966), p. 16, in *Intern. Math. Congress, Abstracts of the Sec. 10*, Moscow; [n] (Russian), Bull. Acad. Sci. Georg. SSR **21** (1966), p. 261; MR **34**, p. 1515; [o] Soviet Math. Dokl. **13** (1972), p. 550; [p] Soviet Math. Dokl. **18** (1977), p. 201; [q] *Problems of Metric Theory of Mapping and its Applications* (Russian), Naukova Dumka, Kiev (1978), p. 109; [r] (Russian), Physica **11D** (1984), 90–108; [s] Lett. Math. Math. Phys. **7** (1983), p. 3; [t] *Operator Theory and Systems*, OT 19, Birkhäuser, Basel, 1986. Proceedings Workshop Amsterdam, June 4–7; pp. 387–392, pp. 393–398; [u] Annals of the New York Academy of Sci. **452** (1985), 255–277; [v] "The Cauchy problem for Heisenberg ferromagnets equation on a finite interval," (preprint IFTT AN USSR, Russian, 1985); [w] "Reflection finite-gap Potentials of the Schrödinger equation and Riemann surfaces" (preprint IFTT AN USSR, Russian, 1987).

Rofe-Beketov, F., S., [a] Dokl. Akad. Nauk, USSR **148** (1963).

Röhrl, H., [a] Bull. Amer. Math. Soc. **68** (1962), p. 125; [b] Comment.. Math. Helv. **38** (1963), p. 84; [c] Math. Ann. **151** (1961), p. 365; [d] Math. Ann. **133** (1957), p. 1.

Serre, J.-P., [a] "Colloque sur les fonctions de plusieurs variables tenu à Bruxelles (1953)," Liège-Paris.

Shiba, M., [a] Hiroshima Math. J. **8** (1978), p. 15.

Sklyanin, E. K., [a] "On complete integrability of the Landau–Lifschitz equation," (preprint LOMI E-3-79), Leningrad (1979).

Spitkovskii, I. M., [a] USSR Sbornik **39** (1981), p. 207; [b] Sov. Math. Dokl. **17** (1976), p. 1733; [c] Sov. Math. Dokl. **22** (1980), p. 471.

Takhtajan, L. A., [a] Phys. Lett. **64A** (1977), p. 235.

Villalon, E., [a] J. of Appl. Math. Phys. **29** (1978).

Volkoviskii, K., L., [a] Sov. Math., Dokl. **16** (1975), p. 1443.

Volkoviskii, L. I., [a] "Modern problems of the theory of functions of complex variables," (Russian), Fizmatgiz, Moscow (1960).

Zakharov, V. E., Manakov, S. V., [a] Funct. Anal. Appl. **19** (1985), p. 11.

Zakharov, V. E., Mikhailov, A. V., [a] Funct. Anal. Appl. **17** (1983), p. 117.

Zakharov, V. E., Takhtajan, L. A., [a] Theor. Math. Phys. **38** (1979), p. 20.

Zakharov, V. E., Shabat, A. B., [a] Funct. Anal. Appl. **8** (1974), p. 43; [b] Funct. Anal. Appl. **13** (1979), p. 13.

Yoshida, M., [a], J. Sci. Hiroshima Univ. S. A1 **32** (1968), p. 181.

SUBJECT INDEX

Abel theorem 35, 76, 114
Abel theorem for generalized analytic functions 148
Abelian differentials, integrals 8–10
Atlas 1

Bers–Vekua representations 131, 134

Canonical bundle K 61
Canonical homology basis 3
Canonical solution of Riemann problem 84
Carleman–Bers–Vekua (CBV) system 131
Cartan–Serre theorem 96
Cauchy kernels 21, 24, 127
Cauchy type integrals 22, 23, 127
Characteristic Chern classes 61, 104
Closed differentials 5, 6
Coclosed differentials 6, 7
Cochains, cocycles 54
Cohomology group 54
Complex line bundles 60
Complex structure 1
Coordinate neighborhood 1
Cutting Riemann surface $\widetilde{M}$ 3
Cyclic section 2

D-BAR method 170
D-cohomologies 101
D-divisor 105
D-divisor degree 105
De Rham theorem 56
De Rham cohomology group 57
Differential form 5, 54
Dirichlet integral 108
Divisor 18, 59
D-Jacobian 122
Dolbeault theorem 58, 61
Dressing method 173

Euler characteristic 4
Exact differential 5, 7
Extremal length 98

Fast-decreasing potentials 149, 165
Fermion field 175
Fine sheaf 56
*-form 6
Fundamental solution of Riemann problem 83

Gardner–Green–Kruskal–Miura (GGKM) equation 154
Generalized analytic functions 131, 132
Generalized constants 138
Green's formula 13

Harmonic differentials 6, 7
Hodge–Royden theorem 7, 101
Holomorphic vector bundle 59
Homology (Betti) group $H_1(M)$ 3
Hyperelliptic surfaces 176

Infinite divisors 107
Inversion of Abelian integrals 35, 39, 43, 125
Its–Matveev formula 164

Jacobi inversion problem 39, 80
Jacobi variety 40
Jost functions 151, 165

Kortweg de Vries (KdV) equations 150

Landau–Lifschitz equation 165
Lax equation 149
Liouville theorem 7
Local coordinate (parameter) 1

Multiplicative constants 145

Non-homogeneous Riemann problem 32, 81

Order of differential 16
Order of function 18
Outer(exterior) product 6

Period of differential 5, 7
Plemelj–Sokhotsky formulae 22, 24, 127

Relationships of neighborhoods 1
Reflection coefficient 151
Reflection finite-zones potentials 156, 168
Reflectionless potentials 154, 161
Residues theorem 8
Riemann bilinear relations 13, 98

Riemann boundary problem 26, 27
Riemann boundary problem, Cauchy index 28, 118
Riemann boundary problem, conjugate 27, 89
Riemann boundary problem, explicit formulae 48, 126
Riemann boundary problem, matrix 82, 178
Riemann boundary problem for generalized analytic functions 144
Riemann–Hilbert boundary problem 183
Riemann–Hilbert problem for differential equations 173
Riemann–Roch theorem 18, 65, 92
Riemann–Roch theorem for generalized analytic functions 138, 143
Riemann surface 1
Riemann theta-functions 42
Riemann theorem 9, 62

Scattering amplitude 151
Scattering data 153, 171
Schrödinger equation 149
Second Cousin problem 72, 97, 111
Serre duality theorem 61, 104
S-divisors 109
Sheafs of germs 54, 59, 61
Short exact sequence 53
Singular group 103
S-Jacobian 122
Soliton solution 154
Special covering 107
S-problems 122

Theta-function 41
Transition matrices of bundles 59
Transmitted coefficient 151

Weyl lemma 7
Weyl–Titmarsh function 156, 157

Zakharov–Shabat equation 160
Zero boundary 98

9 789027 726537